Lecture Notes from the F
on Computer Algebra

Centro Brasileiro de Pesquisas Físicas
Rio de Janeiro, July–August 1989

Editors

Marcelo J. Rebouças
Waldir L. Roque

Available in this series

Volume 1: *Algebraic computing with REDUCE*
Volume 2: *Algebraic computing in general relativity*
Volume 3: *Tackling mathematical problems with MAPLE*

Algebraic Computing with REDUCE

Lecture Notes from the First Brazilian School on Computer Algebra

Malcolm A. H. MacCallum

and

Francis J. Wright

School of Mathematical Sciences
Queen Mary and Westfield College
University of London

CLARENDON PRESS · OXFORD

This book has been printed digitally and produced in a standard specification in order to ensure its continuing availability

OXFORD
UNIVERSITY PRESS

Great Clarendon Street, Oxford OX2 6DP

Oxford University Press is a department of the University of Oxford.
It furthers the University's objective of excellence in research, scholarship, and education by publishing worldwide in

Oxford New York

Auckland Cape Town Dar es Salaam Hong Kong Karachi
Kuala Lumpur Madrid Melbourne Mexico City Nairobi
New Delhi Shanghai Taipei Toronto

With offices in

Argentina Austria Brazil Chile Czech Republic France Greece
Guatemala Hungary Italy Japan South Korea Poland Portugal
Singapore Switzerland Thailand Turkey Ukraine Vietnam

Published in the United States
by Oxford University Press Inc., New York

Reprinted 2011

ISBN 978-0-19-853443-3

Printed and bound in Great Britain by CPI Antony Rowe,
Chippenham and Eastbourne

Editors' preface

This volume comprises the lectures on REDUCE delivered at the *First Brazilian School on Computer Algebra*, held at Centro Brasileiro de Pesquisas Físicas – CBPF, in Rio de Janeiro, from July 24 to August 11, 1989. The other two volumes from the School contain the lectures on MAPLE, SHEEP, MAPLE Applications to General Relativity, and REDUCE in General Relativity and Poincaré Gauge Theory.

The presence of computers has been increasingly felt in almost all branches of human activity, with far-reaching consequences. In science and engineering, particularly, computers have played such an important role that it seems indisputable that young people should develop a sound knowledge of the techniques needed to utilize them in the most effective way.

When most people think about the use of computers in science and engineering these days, they usually consider the traditional territory of computers' use, namely numeric computation. However, computer software packages exist for performing symbolic computation, which is another broad category of scientific computation. These programs are known as Computer Algebra systems and have been used in many problems of science and engineering since the late sixties.

In the last few years the role played by Computer Algebra in science and education has been reinforced by the increasing power of personal computers and the reduction in their cost. This trend can be expected to continue for the foreseeable future and Computer Algebra will eventually become a commonplace tool for the scientific community. Perhaps before the end of the next decade the idea of making algebraic calculations by hand will seem quite archaic.

Among various different actions, we felt that the best way to stimulate the dissemination and use of Computer Algebra as a scientific tool in Brazil was to provide young students and scientists with the opportunity to learn its basis. This was our major motivation when we conceived the School.

In organizing the meeting we took the most traditional view of a School. This meant courses, exercises and practical tutorial sessions. A Computer Algebra laboratory was set up for the practical sessions. The participants were assumed to have no previous experience with Computer Algebra. Nevertheless, we planned the basic courses to cover from the use of Computer Algebra systems as algebraic calculators to advanced programming topics. The present volume clearly reflects our original concerns. The School was divided into two parts. The first two weeks were devoted to general purpose algebraic computing systems while the last week was concerned with Computer Algebra in General Relativity. During the first part two basic courses on REDUCE and MAPLE were offered whereas in the third week we had courses on REDUCE and MAPLE in General Relativity and on SHEEP. In addition to these five courses we also had a SCRATCHPAD demonstration during a School afternoon and a panel discussion on the emerging role of Computer Algebra in Science and Education.

No event of this magnitude can take place without much hard work nor without financial support. Lists of lecturers, organizers and of the sponsors are on other pages of this volume and grateful acknowledgment is made to them all. We are particularly grateful to Professor Malcolm MacCallum for his wise suggestions and advice during the organization of the School. Our appreciation also goes to Dr. Jan E. Åman for his devoted help, especially during the third week of the School. Thanks are also due to many others who helped us in preparing the School. These include the head of our Department João C. C. dos Anjos, our secretary Iara C. Dutra, our colleagues Carla O. F. Barros, Antonio F. F. Teixeira, Renato Portugal, Alberto F. S. Santoro, Luiz A. Reis, Hélio Motta Filho, and our students Filipe M. Paiva, Joel B. Fonseca Neto and Geraldo D. Flores.

It is our hope that the reader will carry with him a sense of excitement appropriate to the emergence of a new subject, and that this book will be useful in helping the members of the scientific community in their studies and use of REDUCE.

While this text was in the final stages of preparation, we learnt of the tragic assassination of Prof. Basilis Xanthopoulos on 27 November 1990. Basilis was one of the most active participants in the School. His warm and friendly, somewhat larger than life, personality made an immediate impression on everybody there. The proceedings of the School, including this book, are dedicated to his memory.

These notes were carefully prepared by the authors and we think they have done an excellent job. We would like to express again our warmest thanks to them for the hard work they put in on each set of lecture notes.

Marcelo J. Rebouças
Waldir L. Roque
Rio de Janeiro, December 1990

Organizing committee of the School

M. J. Rebouças, Chairman – Centro Brasileiro de Pesquisas Físicas

W. L. Roque – Universidade de Brasília

J. Tiomno – Centro Brasileiro de Pesquisas Físicas

H. Fleming – Universidade de São Paulo

I. D. Soares – Centro Brasileiro de Pesquisas Físicas

A. C. Olinto – Laboratório Nacional de Computação Científica

R. A. Martins – Universidade Estadual de Campinas

Sponsors of the School

Conselho Nacional de Desenvolvimento Científico e Tecnológico

Financiadora de Estudos e Projetos

Fundação Banco do Brasil

Coordenação de Aperfeiçoamento de Pessoal de Nível Superior

Fundação de Amparo à Pesquisa do Estado do Rio de Janeiro

Centro Brasileiro de Pesquisas Físicas

Comissão Nacional de Energia Nuclear

International Centre for Theoretical Physics

Financial support

IBM do Brasil

The British Council

The Canadian Embassy

Fundação de Amparo à Pesquisa do Estado de São Paulo

ELEBRA – Divisão de Computadores

Organização dos Estados Americanos

Centro Latino-Americano de Física

Assossiação Brasileira da Indústria de Computadores e Periféricos

Support

Secretaria Especial de Informática

Sociedade Brasileira de Física

Sociedade Brasileira de Matemática Aplicada e Computacional

Sociedade Brasileira de Química

Sociedade Brasileira de Computação

Sociedade Brasileira de Matemática

Laboratório de Física Experimental de Altas Energias – CBPF

Lecturers at the School

J. S. Devitt – University of Saskatchewan, CANADA

R. D. Jenks – IBM Research Center, USA

M. A. H. MacCallum – Queen Mary and Westfield College, UK

J. D. McCrea – University College, Dublin, IRELAND

R. G. McLenaghan – University of Waterloo, CANADA

J. E. F. Skea – Queen Mary and Westfield College, UK

F. J. Wright – Queen Mary and Westfield College, UK

Assistant lecturers

J. E. Åman – University of Stockholm, SWEDEN

G. J. Fee – University of Waterloo, CANADA

R. Portugal – Centro Brasileiro de Pesquisas Físicas, BRAZIL

M. J. Rebouças – Centro Brasileiro de Pesquisas Físicas, BRAZIL

W. L. Roque – Universidade de Brasília, BRAZIL

R. P. Santos – Centro Brasileiro de Pesquisas Físicas, BRAZIL

Authors' preface

The definitive guides to REDUCE are the REDUCE User's Manual (Hearn 1987) and the source code (which is even more definitive, and is always supplied with REDUCE – one of its advantages). We hope that our text will complement rather than compete with these, and we have consciously tried to organize our presentation differently, and usually to use different examples. The text and exercises are based on courses that we have been teaching at QMW (Queen Mary and Westfield College) since 1985, exercises and solutions from which have occasionally been informally distributed. All of our presentation, examples and exercises are original as far as we are aware, except where sources are specifically acknowledged, and apart from our general debt to the REDUCE system and its documentation.

The exercises throughout the book are not all equally easy; we have so far solved most but not all of them ourselves. Some are based on questions from a University of London GCE Advanced Level mathematics examination paper, or on exercises from QMW first year mathematics courses. One or two are intended more to provoke thought about the issues involved – such as just what the capabilities of REDUCE are – than actually to be solved, and one or two go slightly beyond the scope of this book. We intend to make machine-readable solutions (to appropriate exercises) available, either through the REDUCE Network Library (see the Appendix, §F) or directly from us.

REDUCE is available on a very wide range of computers, and almost everything in this book applies (we believe) to all implementations. On the few occasions when it is necessary to consider a specific implementation we will mainly refer to that on the Atari ST, because this is one of the cheapest machines that will usefully run REDUCE, and it is also the system used in our M.Sc. teaching laboratory at QMW and the main system used for the Brazilian School. The Atari implementation is based on Cambridge Lisp (CULisp), and we will refer to such a REDUCE implementation as "Cambridge REDUCE": the examples in this book were produced using Cambridge REDUCE. The "official" Lisp to support REDUCE is Portable Standard Lisp (PSL), and we point out the significant differences between this and CULisp. PSL is used for the recently released implementation of REDUCE for the more powerful IBM PC and compatible microcomputers using the Intel 80386 processor, which we will refer to as "386-REDUCE". However, wherever possible we relegate system-dependent details to the Appendix.

This book is intended to enhance rather than replace the REDUCE manual, but, in order to assist the reader using the book for reference, we have indexed the REDUCE commands we discuss both as individual entries and collected together. (Note that we do not discuss *all* commands.) A similar approach has been applied to RLISP commands, REDUCE switches, and so on. The commands have been separated into "REDUCE commands" (commands used only in algebraic mode), "REDUCE functions" (symbolic mode commands not in RLISP), "RLISP functions" (additional to those in Standard Lisp) and "Lisp functions" (those functions used in REDUCE which are defined in the Standard Lisp report (Marti *et al.* 1979)). Functions specific to certain underlying Lisps or implementations are also listed. Items are indexed where they first appear and at every other point where their properties are discussed, but not where they are merely used in constructing examples of other features of REDUCE. The table of contents will also be useful to readers aiming to look up a specific topic.

Basilis Xanthopoulos, to whom this book is dedicated, was very well known to one of us (MAHM) since the two had closely related research interests in relativity: perhaps the relationship and Basilis's generous spirit are best illustrated by the fact that he invited MAHM to share a dinner table on the same day as the two had had a strongly-argued public clash at the GR12 conference in 1989. He will be very much missed.

We wish to thank the following: John Fitch and Arthur Norman for providing us with versions of Cambridge Lisp and REDUCE that run on our slightly idiosyncratic choices of computers; Anthony C. Hearn for permission to quote from the REDUCE source code and manual; all three of them for tolerating our questions over a number of years, and even answering most of them; Marcelo Rebouças for catalysing the writing up of our lecture notes and not complaining too much each time we missed a deadline; Adrian Burd, Barbara Davies, Anthony C. Hearn (especially), Grant Keady, Jed Marti and Andreas Strotmann for their perceptive comments on drafts of this book; our wives and families for not reminding us too often that there is more to life than computer algebra.

M. A. H. M. and F. J. W.
QMW, London, January 1991

Contents

Tables and Figures

1

REDUCE as an algebraic calculator

In this book we present almost all of the REDUCE computer algebra system, plus some mathematical and computational topics that go beyond the basic system. Some technical remarks are included in an appendix.

In this chapter we set the scene for algebraic computing, and then specialize to REDUCE, deal with the main technicalities of running it, and show how to use it to perform simple interactive calculations.

1.1 An overview of computer algebra systems

This book is primarily about the algebraic computing system REDUCE, and we will focus on the latest version[1] which is 3.3. But why learn REDUCE particularly? Algebraic computing systems fall into two main classes: general purpose and special purpose. General-purpose systems sacrifice speed in favour of flexibility and ease of use, as can be seen by comparison with a system such as SHEEP[2], a special-purpose system designed for use in general relativity.

Before getting down to detail, it is useful to place REDUCE in context among the main contemporary general-purpose computer algebra systems, which are listed in Table 1.1 in order (very roughly) of decreasing demand on the host computer. This ordering also corresponds to decreasing cost and decreasing power, but only in a very

[1]Specifically, we describe the latest version that we have received from our distributor, which we believe to be the currently *distributed* system – we do not regard as part of this standard system enhancements that are available but not, as far as we are aware, currently distributed, although we mention some of these where appropriate.

[2]Described in volume 2 of this series.

rough sense. The arrows show new systems that have been developed from older ones and have made their predecessors essentially obsolete.

Table 1.1. The main general-purpose systems

System Name (original → successor)	**Implementation Language**
Scratchpad	Lisp
MACSYMA	Lisp
SMP → Mathematica	C
REDUCE	Lisp
Maple	C
muMATH → Derive	Lisp

For a detailed overview of the five general-purpose systems currently available (not Scratchpad) see the book by Harper *et al.* (1991).

The two longest-established algebraic computing systems are MACSYMA and REDUCE, which were first designed in the mid 1960's. They represent two distinct traditions: REDUCE was originally developed to solve problems in high energy physics, whereas MACSYMA was originally developed as an illustration of programming techniques that were then considered to represent artificial intelligence. Both have transcended their origins, and continue to be developed as general-purpose algebraic computing tools in their own right. REDUCE is being intensively developed under the control of its original designer Anthony C. Hearn, although MACSYMA is probably the most extensively developed system in existence. REDUCE was designed with portability in mind, and hence runs on the widest range of computers. Versions of both REDUCE and MACSYMA have recently become available for Intel 80386/MS-DOS-based microcomputers.

Scratchpad II is an exciting IBM project, which allows much more abstract mathematical objects to be constructed and manipulated than do other systems, and should therefore appeal particularly to pure mathematicians. The disadvantage is that it runs only on IBM computers, requires a very powerful machine, and (as far as we are aware at the time of writing) has still not become commercially available. Maple[3] is a similarly well-founded system that is being actively developed at the University of Waterloo in Canada, and was originally designed for teaching. SMP was a rather pragmatically-based system

[3]Maple is a trademark of the University of Waterloo. The system is described in volume 3 of this series.

aimed at physicists and engineers; *Mathematica*[4] appears to retain some of that flavour but provides an extremely powerful and flexible all-purpose scientific calculator and graphical display system.

DERIVE[5] and the most recent versions of muMATH are available *only* on PC-DOS/MS-DOS[6] systems, and muMATH at least cannot make use of more than 256Kb (Kilobytes) of main memory for data. muMATH was written specifically for teaching, and was a pioneering system in this area. It was designed originally to run on 8-bit CP/M[7] systems and then updated to run on the 16-bit IBM-PC and compatibles, which were then the only cheap and readily available microcomputers, but it now appears hampered by their architecture. However, either muMATH or DERIVE (its successor) is probably the best system to run on Intel 8086-based microcomputers.

REDUCE and Maple will run satisfactorily for simple problems in 1Mb (Megabyte) of main memory, and hence can be usefully run on modern cheap but powerful microcomputers such as the Atari ST. *Mathematica* and MACSYMA need something like 5Mb of main memory on a machine such as an Apple Macintosh II, SUN, NeXT, DEC VAX[8] or Symbolics computer.

This introductory sketch shows that there are several good reasons to learn REDUCE. It has a good pedigree and is still being actively developed. It is very flexible, reasonably easy to use, and among all algebra systems runs on probably the widest range of computers, from the IBM PC to the Cray X-MP. Once you know one system it should be quite easy to pick up the details of any other. If we had to choose another system to use instead of REDUCE for our own teaching and research it would probably be Maple, although there are important differences in the style of operation of the two systems which we feel tend to favour REDUCE for many research applications.

1.2 REDUCE in general

From now on we will focus on REDUCE 3.3. There is a price to be paid for using a system that is under active development: in order to provide major improvements it is sometimes necessary to sacrifice some

---

[4] *Mathematica* is a trademark of Wolfram Research, Inc.

[5] DERIVE and muMATH are trademarks of Soft Warehouse, Inc.

[6] PC-DOS is a trademark of IBM; MS-DOS is a trademark of Microsoft Inc.

[7] CP/M is a trademark of Digital Research Inc.

[8] VAX is a trademark of Digital Equipment Corporation (DEC).

compatibility with previous versions. This has happened in a small way with REDUCE 3.3. Versions 3.1 and 3.2 just provided corrections of errors that were discovered in REDUCE 3.0, but version 3.3 has some significant new facilities. Because version 3.3 may not yet be available at all sites we will mention in footnotes the differences between it and previous versions of REDUCE 3.

We will occasionally mention minor aspects of REDUCE 3.3 that we think are wrong or inconvenient, and sometimes we will suggest ways around them. We are assured by Anthony Hearn that many of them will be corrected or improved in future versions of REDUCE, and we will mention one or two developments that we know are planned. However, we believe that the changes likely to be made in the next version will not be as significant (at least for users) as those between versions 3.2 and 3.3; time may prove us wrong.

Information about availability of REDUCE for different computers may be obtained from the RAND Corporation at the address given in the bibliography under Hearn (1987); for electronic mail access see the Appendix, §F.

REDUCE is written in RLISP, which is a language based on ALGOL 60 syntax that has a 1-to-1 correspondence (similar to that between assembly language and machine code) with a dialect of Lisp called Standard Lisp. This is not standard in the sense that FORTRAN 77 is standard, but it is a small Lisp designed to be sufficient to support computer algebra. It is available as a subset of most other Lisps, possibly after minor modifications of a few functions. This and the fact that REDUCE requires very few facilities from the computer operating system are the reasons why REDUCE is available on so many systems. The penalty for this machine independence is that the distributed REDUCE system does not at present contain facilities that require much system-dependent code, such as graphics and sophisticated editing. However, packages are under development to fill some of these gaps, possibly in a system-dependent way.

A criticism of Lisp in the early days was that it was slow, because it is basically an interpreted language. However, modern professional Lisp systems contain a compiler, and there is no reason why compiled Lisp should run significantly slower than FORTRAN on the same problem. (There is at least one Lisp system that was written in FORTRAN.) But Lisp is vastly more flexible, as required for computer algebra, and when using that flexibility it may run more slowly. All of the REDUCE system itself is compiled, and in most implementations it is very easy for users to compile their own REDUCE procedures –

on many systems this is the default[9]. REDUCE may well be faster for some purposes than systems not written in Lisp, such as SMP and Maple, that do not (as far as we are aware) provide a compiler for the user's programs.

Since the rest of this book will be about what REDUCE *can* do, we should mention briefly a few things that it *cannot* at present do, or cannot do well: it does not provide definite integration (unless the integrand is indefinitely integrable) or higher transcendental ("special") functions, and graphics and numerical facilities are limited compared with some other systems.

For alternatives to our own point of view there is a short article by Fitch (1985) that introduces REDUCE 3.1 through some interesting examples, and we are aware of (only) three other books in English that describe REDUCE (we include a number in other languages – principally Russian and Japanese – in the bibliography). The book by Rayna (1987) is the only one devoted entirely to REDUCE; it is based on REDUCE 3.2, but much of it is relevant to REDUCE 3.3 and we refer to it occasionally for details of particular topics. Davenport *et al.* (1988) present a summary of REDUCE 3.3 in a 20-page annex, and refer to REDUCE and other systems at appropriate points throughout their book. The lectures by Winkelmann and Hehl (1988) are the closest to our own in spirit and describe REDUCE 3.3. However, none of these texts considers REDUCE symbolic-mode programming, and only Davenport *et al.* (1988) discuss internal representations and algorithms. The book by Buchberger *et al.* (1983) is currently the definitive theoretical survey of computer algebra, and is essential background reading for those interested in the theoretical aspects, but it is not an easy book and the interested reader may be well advised to proceed from our text via Davenport *et al.* (1988) to Buchberger *et al.* (1983).

1.3 Interactive REDUCE

Like all the other algebraic computing systems mentioned above, REDUCE is normally used interactively, although it can also be used in "batch" or "background" mode (see §3.10). A small number of details

[9]On certain implementations, especially those for PC-DOS/MS-DOS machines based on Intel 8086 and 80286 (but not 80386) processors, it is effectively impossible to compile in new versions of REDUCE's internal functions, with the result that, for example, certain programs in the REDUCE network library (see the Appendix, §F) will not work in these versions.

of REDUCE usage depend on the computing system on which it is run, and how to start it is one such detail.

1.3.1 Starting REDUCE

There are two main user interfaces in current use: command line and graphical. With a properly configured command-line interface, such as that provided by MS-DOS or a standard UNIX[10] *shell*, typing the command "`reduce`" after the command prompt should start REDUCE running, although it may be possible to exercise additional control by giving command-line arguments. With a graphical interface, a program is started by pointing, usually using a *mouse*, at either the name or a picture of the program (an *icon*) and *opening* it, often by pressing a mouse button twice quickly (*double-clicking*).

A system may provide several alternative ways to run REDUCE, and the precise setup is at the discretion of the user or the system manager. The Atari ST supports various command-line and graphical interfaces, and we give a few more details in the Appendix, §A.

If it is not possible to start REDUCE by simply typing "`reduce`" or double-clicking on an icon, it will be necessary to use a more explicit invocation. In order to explain how to run REDUCE the hard way it is useful to take a quick look at its internal structure – in particular it is important to appreciate that REDUCE operates "on top" of Lisp. Lisp is a highly dynamic computing language, quite unlike such conventional languages as FORTRAN, Pascal and C, and more like interactive BASIC, and the Lisp system must always be there "underneath" a system written in Lisp. The core of a Lisp system is usually written in some conventional system-programming language, which is BCPL in the case of Cambridge Lisp (CULisp).

With CULisp, for example, this core is run first and dynamically loads the rest of the system, which is written in Lisp itself, as a "memory image". To run REDUCE it is necessary to load the right image file, which is usually achieved by running Lisp with suitable arguments. With a command-line interface it would probably be necessary to type something like

```
% lisp -image /reduce/image
```

where "`%`" represents the command prompt. But the precise mechanism is, of course, system dependent, and we give another example relating to the Atari ST in the Appendix, §A.

---

[10]UNIX is a trademark of AT&T Bell Laboratories.

With PSL, by contrast, most of the system is normally built into a much larger initial executable program file and much less is dynamically loaded. This difference seems to affect the initial startup time for REDUCE, which in our experience is slower for Cambridge REDUCE than for PSL-based REDUCE.

Where appropriate in this book we will use an italic typeface thus: "*description*", to *describe* either what REDUCE will output or what the user should input, as opposed to showing a specific example. The first instance of this follows.

REDUCE is large, hence it *may* take a while to load, but once loaded it will announce itself something like this:

System information
`REDUCE 3.3` *Release date*

```
1:
```

REDUCE is now waiting for input. All REDUCE input prompts take the form of a number followed by a colon (and a space). The number can be used to refer back to previous input and output as we will see later. To convince yourself that REDUCE really is alive try typing some simple input, such as

```
1: X;
```

where the semicolon must be followed by pressing the *Return* (or *Enter*) key. There *may* be a long pause, possibly longer than it took for REDUCE to load initially, followed by the exciting response

```
x
```

One can learn quite a lot by analysing what is going on here. REDUCE has interpreted the input as a statement asking it to output the current value of the variable called `X`, which has not yet been given a value so it remains symbolic. In other words, the current value of `X` is `X` itself, which is what REDUCE has output. This is one essential feature of an *algebraic* computing system. It may take REDUCE a while to do this, partly because it has to go through a lot of its machinery for interpreting a general statement, and partly because (at least in some implementations) REDUCE is composed of a very large number of small modules, which may be loaded only as needed. REDUCE does not need many modules to display its initial prompt, but it needs a lot more to respond to any input statement.

In some implementations, particularly those based on CULisp, REDUCE (actually the underlying Lisp) keeps modules loaded unless it starts to run short of memory, and then it starts to discard them, probably on a first-in first-out basis. This usually spells disaster with REDUCE, because it implies that there is not enough memory to solve the problem. REDUCE may immediately need to reload the very module it has just discarded, a cycle which is referred to as "thrashing". A lot of disc activity with no output after a long time suggests that this is what is happening. In Cambridge REDUCE the statement

```
LISP VERBOS 3;
```

turns on a display of information about subsequent internal activity (but it must be given *in advance*).

1.3.2 Terminators

The other important feature of the toy input above is the semicolon. Every statement or command to REDUCE must end with a *terminator*, such as a semicolon, and REDUCE will not do anything with a statement or command until it gets a terminator. If you forget, you can find yourself waiting a long time for REDUCE to respond, when all the time REDUCE was waiting for you! The terminator can be on the same line as the statement, or on a subsequent line. What happens is that REDUCE *reads* each line of input when the Return key is pressed. It does not read individual characters as they are typed; these are processed by the computer's operating system, so that whatever input editing facilities are provided by the operating system can be used in REDUCE. When REDUCE has read a line of input it processes all the statements up to the last terminator, and any incomplete statement is remembered until it is completed on a subsequent line. Hence the input is completely free-format, as in Pascal and C, and unlike current versions of FORTRAN, so that there can be several statements on one line, or one statement can span several lines. In summary, a terminator followed by *Return* is required to get input to REDUCE read and processed.

The algebraic value of a statement at the "top-level" of statement nesting[11] that is terminated with a semicolon ";" will automatically be output. If a statement is terminated with a dollar sign "$" then its value is not output, which is essential in cases where the output is known to be very long (as is often the case in algebraic computing) and not interesting. An algebra system can spend longer preparing a value for output than computing it in the first place!

[11] Precisely what this means will be made clearer in chapter 3.

1.3.3 Stopping REDUCE

The way to stop REDUCE tidily at the end of a session is to type the command

```
BYE;
```

(including, of course, the terminator and *Return*). There are other ways with system-dependent effects: "`QUIT;`" is often the same as "`BYE;`", but on some systems it stops REDUCE in such a way that it can be restarted at the point where it was stopped. Some operating systems, such as some versions of UNIX, provide a way of stopping any program in a restartable way, and this can be used if available. It can be very useful to switch between REDUCE and a file editor in this way, since standard REDUCE does not provide file editing. (Alternatively, it is possible to run a program such as REDUCE from within some editors[12], and even to capture the REDUCE input and output into an edit buffer.)

Many Lisp systems provide a way to interrupt execution, which may be necessary if a REDUCE calculation runs out of control, or just takes longer than expected. Often the interrupt character will be *Control-c* (hold down the *Control* key and type `c` or `C`), which should cause a "Lisp break" possibly allowing several options, such as aborting or quitting the current REDUCE statement or command and returning to a new REDUCE input prompt, continuing the current execution, or stopping REDUCE completely. It is worth finding out how these facilities work while not doing anything important.

1.4 Comments and case sensitivity

By far the most widely used comment convention in modern REDUCE is that everything following a `%` character until the end of a line is regarded as a comment for human consumption, and is read but otherwise ignored by REDUCE. The `%` can appear at the start of a line (like the C or `*` in FORTRAN 77), or after part of a statement (like the `!` in FORTRAN 90). In the latter case it is essential that any terminator for the preceding statement appears before the `%`, because otherwise it will be regarded as part of the comment and ignored. Since such a comment is terminated by the end of the line, any continuation of the comment on a following line must begin with its own `%`.

REDUCE also supports the ALGOL-style comment convention that any statement beginning with the keyword `COMMENT` and ending with

[12]E.g. GNU-Emacs by the Free Software Foundation.

a statement terminator is a comment. This comment statement is a genuine statement and follows the same rules as any other statement; it may be one of a number of statements on one line or may extend over more than one line. Such a comment statement may appear *anywhere* that a blank or space could syntactically appear without affecting the meaning of other statements, and is the only form of comment that can be followed by program code on the same line. It provides a convenient way of writing a multi-line comment that can reliably be re-formatted by a line-wrapping editor. It also provides a mechanism for commenting out parts of the code on a statement rather than a line oriented basis, but it only works for simple statements: any statement that does not contain any block constructs (i.e. does not include either `<<` or `BEGIN` – see §3.7) can be commented out by preceding it with the keyword `COMMENT` and using the statement's own terminator to terminate the resulting comment statement.

By default, REDUCE input is case independent, like most modern implementations of FORTRAN. Most REDUCE implementations convert all input to upper case and produce all output in upper case, although Cambridge REDUCE converts to lower case. Case insensitivity has the advantage that you cannot get the case wrong, but the disadvantage that you cannot use the same letter in upper and lower case to represent two different things. Using mixed case can make programs more readable; for example, the meaning of the variable `SinTheta` is probably clearer than if it were written as `sintheta`, although in REDUCE they would be equivalent. (It is possible to make REDUCE case sensitive, but this is probably inadvisable because, for one thing, it makes code less portable. It is also possible to force letters to retain their case by preceding them with the so-called "escape character", the exclamation mark "`!`", to which we return in §1.6. This can be useful for keeping REDUCE output close to a particular mathematical notation, but the input looks a little ugly so we will not use this technique here.)

However, for purposes of demonstration we will use alphabetic case differently. In our examples, it would be confusing to show prompts because the actual prompt numbers are not significant, and also it is harder (in most REDUCE implementations) to capture a complete session including prompts as well as input and output. Therefore, we will show example input without prompts but normally in UPPER CASE, whereas the output will be in lower case since we are using Cambridge REDUCE. We will use this convention only where there would be ambiguity otherwise, and normally not in comments or longer code examples such as procedure definitions. We will also use a typewriter-style font for all REDUCE code: e.g. `THIS IS TYPEWRITER-STYLE TEXT.`

Our REDUCE output is all captured from actual runs of Cambridge REDUCE, and in presenting it we have tried to preserve the general appearance of REDUCE output to a screen or printer, although sometimes it is slightly edited – in particular, some blank lines have been edited out to save space.

1.5 Computer calculus? Differentiation and integration

It has been suggested that "computer algebra system" is a bad name because it disguises the fact that these systems can apparently perform *analysis*, or infinitesimal calculus, as well as algebra. But, in fact, they can only perform calculus algebraically and *not* "from first principles", and they cannot perform the proofs of analysis on which the methods of calculus depend. Of course, in practice human mathematicians also usually perform calculus algebraically, because it is much quicker. But a human mathematician could deduce the derivative of some new function by going back to basics, and for the pathological functions that lecturers like to use to demonstrate non-differentiability the simple algebraic rules cannot be used. In such cases a computer algebra system would probably fail, and to emulate what a human would do would require much more artificial intelligence than current computer algebra systems possess. (There is a research project lurking in this observation.)

For the well-behaved smooth functions that normally arise in science and engineering, computer algebra systems can effectively perform differentiation and integration of restricted classes of functions. A system can only differentiate functions whose derivatives it has been told, and combinations of those functions, but it is trivial to add new differentiation rules. The restrictions on indefinite integration are more fundamental and it is harder to add the capability to integrate a new class of functions, but as we will see in more detail in chapter 7 it is possible to integrate algorithmically a class of functions including rational functions and functions involving exponentials and logarithms (which includes trigonometric and hyperbolic functions), and quite recently progress has been made with the integration of functions involving square roots.

Definite integration is even worse because there is no *general* theory, and systems (like MACSYMA) that provide sophisticated definite integration facilities usually use an *ad hoc* collection of techniques such as contour integration and table look-up. REDUCE currently has no definite integration facility, although as we will see it is trivial to integrate definitely a function that REDUCE can integrate indefinitely (although a naïve approach may not give correct results in the presence of singularities).

The REDUCE differentiation operator is called DF. As an example, let us differentiate a function that is simple to specify but tricky to differentiate by hand: x^{x^x}. The REDUCE statement to do this is

```
DF(X**(X**X), X);
```

and the result[13] (in lower case because we are using Cambridge REDUCE) is

```
   x
  x  + x        2
 x       *(log(x) *x + log(x)*x + 1)
------------------------------------
                 x
```

This conveniently provides a non-trivial function to integrate indefinitely, and to save re-typing REDUCE provides the predefined variable WS (which stands for WorkSpace) as a way to refer back to the last algebraic value computed. In this and all subsequent examples we assume that it will be clear without annotation which is input to REDUCE and which is output:

```
INT(WS, X);

  x
 x
x
```

In the above two examples it is fairly obvious that the first argument of the operator DF or INT is the expression to be differentiated or integrated and the second argument is the variable with respect to which the operation is to be performed. This is the only way that INT can be used, but DF has more general forms. The operator DF is really a partial differentiation operator, and it can differentiate with respect to an arbitrary number of independent variables in one go, e.g. the following computes $\partial(xz\sin y + a(\log(xz))/y)/\partial x\partial y\partial z$:

```
DF(X*Z*SIN(Y) + A*LOG(X*Z)/Y, X, Y, Z);

cos(y)
```

Moreover, if any independent variable is followed by an integer then it means differentiate that many times with respect to that variable, e.g. the following computes $\partial^4(kx^3y^2)/\partial x^2\partial y^2$:

[13] The "vertical" output of rational quantities is new in REDUCE 3.3 – previous versions used only "/".

```
DF(K*X**3*Y**2, X, 2, Y, 2);

12*k*x
```

This feature has the consequence that the integers denoting the order of differentiation must be explicit numbers and cannot be purely symbolic, because otherwise REDUCE would think that they were independent variables. There are ways around this, as we will see in §4.11.

Having introduced two REDUCE operators, `DF` and `INT`, it is worth stressing at this point that, in common with most (but not all) other REDUCE operators, they fully evaluate *all* their arguments before they use them. Hence, whenever we state that an operator argument must be a particular type of object, what we really mean is that it can also be an expression – which will often be simply a variable with an assigned value – that evaluates to the required type of object. Hence, the following rather silly example correctly evaluates $d^2(x^3)/dx^2$:

```
DF(X**3, (1/X**3)**(-1/3), SQRT 4);

6*x
```

Arguments to `DF` after the first must evaluate to variables optionally followed by non-negative integers, and the second argument to `INT` must evaluate to a variable.

If you ask REDUCE to differentiate an expression with respect to a variable on which it does not depend, then REDUCE will quite rightly return the result as zero:

```
DF(Y, X);

0
```

But one often manipulates a derivative such as dy/dx by hand as a symbolic object, e.g. when defining a differential equation. Such a symbolic derivative implies that y depends on x, but in some unspecified way. Unspecified dependence is established in REDUCE by the declaration `DEPEND` (which does not end with an s – think of it as short for *dependence* rather than *depends*), e.g.

```
DEPEND Y, X;
DF(Y, X);

df(y,x)
```

declares y to depend on x and demonstrates that $\partial y/\partial x$ then remains symbolic.

The `DEPEND` command also works for other linear operators (of which DF is just a special case), as we will see in §4.8.4. Dependence can be removed by the declaration NODEPEND, which has exactly the same form as DEPEND. Both DEPEND and NODEPEND allow an arbitrary number of *independent* variables to be specified, e.g.

```
DEPEND F, X, Y, Z;
```

declares F to depend on each of X, Y and Z, and

```
NODEPEND F, X, Z;
```

removes the dependence on X and Z leaving only that on Y.

1.6 Variables, constants and expressions

1.6.1 Variables and identifiers

We have informally introduced variables, constants and expressions already, hoping that the way they work is so obvious that it needs little explanation. The variables used so far have all been identified by single letters, as is common in mathematics. However, in computing it is usual to use longer names that *describe* the nature of the variables. This is possible in REDUCE, which allows identifiers to be sequences of at least 24 characters, and many implementations allow much longer identifiers. The only constraints are that if any character other than a letter or digit is to be included it must be *escaped*, meaning that it is preceded by the "escape" character "!" (which itself is not regarded internally as part of the identifier) and if the first character is anything other than a letter then it must be escaped.

It is quite common to include escaped hyphens and underlines in identifiers to make them more readable (and more likely to be unique), and it is possible, although probably not advisable, to include spaces similarly. Identifiers used internally in REDUCE often contain characters such as *, :, =, so these should not be included in your own identifiers. Some examples of acceptable REDUCE identifiers, and the way that Cambridge REDUCE outputs them, are:

```
{X0, ALPHA, !Very!Long!Variable!Name, Make!_Hilbert,
Hessian!-Matrix, S! P! A! C! E! D! ! ! O! U! T};

{x0,

 alpha,

 VeryLongVariableName,

 make_hilbert,
```

```
hessian-matrix,

s p a c e d   o u t}
```

For simple interactive use of REDUCE one tends to use short variable names to avoid typing, but long descriptive names become very useful when writing large programs in REDUCE. Identifiers are used to identify not only variables but also all other objects in REDUCE except explicit constants.

1.6.2 Number representations

At this point we need to fix our terminology. Although the integers form an infinite set it is countable, so that a subset can be represented exactly in an explicit and finite way. Hence integers can be represented in a computer in essentially the same way that one would write them on paper, the only difference being that usually the radix (or base) used is 2 rather than 10. The rationals (or rational numbers) also form an infinite but countable set, and a finite subset of them can be represented as pairs of integers. However, the reals (or real numbers) form a set that is not only infinite but also uncountable; it is a continuous set, which may be considered to include the rationals and integers as countable subsets. An irrational real number has no exact representation that is both explicit and finite, e.g. the representation of the circular constant as π is finite but implicit, whereas 3.14159 is explicit and finite but is only an *approximation*.

In computing terminology, the terms *integer*, *rational* and *real* mean those subsets of the infinite mathematical sets that are representable in the particular computing system being used, which will depend on both the hardware and the software. This is the meaning that we will ascribe to these terms in future except when we are discussing mathematics, which we hope will be clear from the context. In order to be representable, computer real numbers are a rational subset of the mathematical reals, but in order to capture better the nature of the real numbers the representation used for reals is different from that used for rationals or integers. *Real* normally implies *floating point* representation, which is sometimes referred to as "scientific" representation. It approximates the number to a fixed *relative accuracy* by storing a fixed number of *significant* digits in a component of the representation called the *mantissa* and storing the overall magnitude in the form of a power of some fixed base in a component called the *exponent*. (A fixed point representation could also be used, but in practice it rarely is except for integer representation, which can be thought of as a special case.)

Conventional languages such as FORTRAN, C and Pascal have essentially one representation for integers and one for reals, although there is some small choice in the amount of storage allocated to the representations and hence in their overall range and/or precision. REDUCE differs by providing a representation for integers that is restricted only by the amount of memory available, a representation for rationals as quotients of integers, a representation for reals (called `float mode`) that is similar to that used in conventional languages and usually uses a base of 2, and also a much more flexible representation[14] (called `bigfloat mode`) that uses arbitrary-size integers to represent both the mantissa and the magnitude as an exponent of 10. (Most other algebraic languages provide similarly flexible numerical representations.)

Thus in algebra systems the rational numbers exist as a primitive data type, rather than only as expressions to be evaluated (perhaps approximately using floating-point representation). The important distinction between rational and real number representations in a computer is that, within their allowed ranges, rational implies an exact representation whereas real implies an approximation. Because algebra systems are primarily used for exact manipulation of algebraic expressions, arbitrary-size integer and arbitrary-precision rational representations are more fundamental than real representation, and so will normally be used by default.

1.6.3 Constants

REDUCE provides the same formats for inputting integer and real constants as do most other programming languages, and numerical constants can be signed. For example, `-12345` represents an integer constant, whereas `1.234E5` represents the real constant 1.234×10^5. Entering it in this form would normally imply that it was being regarded as an approximation to a real number rather than as a rational number, and that an appropriate numerical domain had been selected – see §1.10. Using "E-format" as a shorthand for a rational number is a bad idea at present, because the way that REDUCE converts reals to rationals may be unreliable, especially for very large and very small reals (partly due to inherent inaccuracies in fixed-precision floating-point representation; see also §A.3). But an important difference between REDUCE and conventional languages is that integer constants can have arbitrarily many digits, and in `bigfloat` mode (see §1.10) so

[14] We expect future versions of REDUCE to provide also a "rounded" mode that combines the two current real modes.

can real constants, so that they can be arbitrarily large and arbitrarily accurate. (Although arbitrary here should be regarded as a technical term meaning within memory limits, and for input not extending over more than one line in current standard implementations of REDUCE.)

REDUCE also provides (as do all other algebraic computing systems) a format for specifying rational constants, which by default are stored in exact rational representation. A rational constant is input as a quotient of two integers (which must be co-prime and the divisor must be positive, otherwise it is a rational expression that requires simplifying) and since these two integers can have arbitrarily many digits the rationals can approach arbitrarily close to any real number. For example, 1234/56789 is a rational constant.

The only other type of explicit constant in REDUCE is the string, which is an arbitrary string of characters in double quotes, thus: `"This is a string"`. These are used primarily for outputting messages to the user, although in principle they have the same status as other types of constant – there are just not many operations defined on them. REDUCE (at the algebraic level) has really no facilities for manipulating character data, and it does not even have the flexibility to include a "`"`" character within a string. To do this kind of thing it is necessary to work in the symbolic or RLISP mode, which will be described in chapter 6.

1.6.4 Expressions and functions

Algebraic expressions are combinations of variables and constants with arithmetic operators and function calls exactly as in most other languages, and simple examples have already been demonstrated. The arithmetic operators are `+` (addition), `-` (subtraction), `*` (multiplication), `/` (division) and `**` (exponentiation). Exponentiation is the operation of "raising to a power" which is available as an infix operator in FORTRAN, but not in Pascal, Modula or C. In many, if not most, modern versions of REDUCE the symbol ∧ is available as a synonym for `**`, and is the symbol used in most BASICs that provide exponentiation. However, this use of ∧ is not completely standard and the symbol is required for other purposes, such as vector or exterior (wedge) product, in some REDUCE packages, so it is safer to avoid it for exponentiation.

The behaviour of these operators is standard, and follows the normal rules of arithmetic. Parentheses (round brackets) can be used as usual, so if there is any doubt about the interpretation of an expression it is best to use parentheses to make it unambiguous. For example, the rational expression

$$\frac{a+b}{c+d}$$

would be input to REDUCE as `(A+B)/(C+D)`. Two common ambiguous cases are `A/B/C` and `A**B**C`, which REDUCE interprets as `(A/B)/C` and `(A**B)**C` respectively, but rather than trying to remember this it is easier to use parentheses always.

Functions are referenced exactly as in other languages, e.g.

```
A * COS( X*(A+Y) + EXP((X**2)/2) );

               2
              x
cos(sqrt(e)     + a*x + x*y)*a
```

(where `sqrt` is the square root function). Note that the REDUCE output is slightly different from the input, but mathematically equivalent. REDUCE knows a little about standard functions such as `COS` and `EXP`, but if you reference a function that REDUCE does not know about then it will ask you (if running interactively) whether you want to declare it an operator. Type `Y` (or `y`) in response to this query, unless the new function was in fact a typing mistake, in which case answer `N` (or `n`) and correct the mistake. (We will see in §2.2 how to edit previous input.) REDUCE is quite happy for you to use functions for which no properties have been defined, although it cannot do anything at all with them, and in fact it treats them essentially like variables with "compound" names (in fact they are "kernels" in REDUCE: see §2.9 and §5.3).

1.6.5 Symbolic constants

REDUCE has a facility for defining named constants as do most other languages. The REDUCE facility is implemented as a parameterless macro like a restricted version of `#define` in C, or a less restricted version of the `PARAMETER` statement in FORTRAN 77 (to which we return in §9.3), rather than the `const` declaration available in Pascal and ANSI C. The syntax is

`DEFINE` *name1* `=` *const1*`,` *name2* `=` *const2*`, ...`

for example,

```
DEFINE THIRD = 1/3,  CR3 = 3**THIRD;
DEFINE ZILLION = 1000000000000;
```

The `DEFINE` command defines a purely textual replacement that is implemented when REDUCE reads its input, so that inputting a `DEFINE`d identifier is exactly equivalent to inputting its definition. This is true within DEFINE statements themselves, so that DEFINEd identifiers can be used in subsequent definitions, as above in the example that defined CR3 to be the cube root of 3. But unfortunately, in the current implementation of DEFINE an identifier on the left of the = sign is also replaced by any defined value it may have, making it impossible to redefine an identifier, and no facility is provided to remove or clear such definitions[15], so that DEFINE must be used with care.

The REDUCE DEFINE command is more general than the FORTRAN 77 PARAMETER statement in that it allows completely general expressions including function calls as the defined values. (In fact, the example above already illustrates this, because the FORTRAN 77 PARAMETER statement allows ** only with integer exponents.) Here are some more examples:

```
DEFINE PIBY2 = PI/2,  GOLD = (1+SQRT(5))/2,
       COSINE = COS,  LN = LOG;
```

So when might one use DEFINE in preference to the related but more algebraic facilities provided by assignment and let-rules, which we will introduce later? A DEFINEd identifier used as a named constant cannot be accidentally (or intentionally!) changed (although the side-effects of the attempt may be undesirable), and other attempts to misuse such a named constant are likely to generate syntax errors. A DEFINEd identifier will be replaced by its definition regardless of where it appears, whereas in a few situations – principally on the left of an assignment or let-rule – a bound variable is not replaced by its assigned value.

Algebraic computing also needs another kind of named constant that has no analogue in conventional languages; this is the purely symbolic named constant that represents itself rather than a numerical (or string) value. There is no general mechanism in REDUCE for distinguishing these from variables, and any distinction remains purely in the mind of the user. You would probably naturally regard

$$ax^2 + bx + c$$

as a polynomial in x with constant but unspecified coefficients a, b and c, but it would be just as valid to regard it as a linear polynomial in the variable a, say, with coefficients that have slightly eccentric forms. By default, REDUCE would tend to take the latter point of view, which is why it is nearly always necessary to specify what the variables are

---

[15]The DEFINEd value of the identifier NAME can be removed with the statement "REMPROP('NAME, 'NEWNAM);", but this uses facilities from symbolic mode (which is described in chapter 6). Similarly, identifiers can be effectively reDEFINEd using symbolic mode.

when any operation is performed in which it matters, such as finding the degree of a polynomial.

1.7 Algebraic efficiency

As a very simple illustration of the general situation, note that X**2 is an expression that REDUCE cannot simplify further, whereas X*X will cause REDUCE to perform a multiplication to turn it into X**2. Pascal users are used to writing X*X because an exponentiation operator is not provided in Pascal, and many FORTRAN users have the misguided notion that X*X is "more efficient" than X**2 (which may have been true in days long gone by before people knew how to write compilers properly). However, there are big differences between "numerical efficiency" and "algebraic efficiency", and in some cases what is most efficient algebraically would be least efficient numerically. Exponentiating variables is one such trivial example – remember that in an algebra system X**2 is a fixed polynomial, whereas X*X is an expression to be evaluated. Of course, avoiding unnecessary re-evaluations is essential for efficiency in any language, but in general determining the most efficient algebraic implementation of an algorithm is still a research problem (Pearce and Hicks 1981; Davenport *et al.* 1988).

1.8 Assignments and equations

So far, we have used variables as mathematical "unknowns", and such variables are called *free* or *unbound*. Conventional languages (including Lisp) do not allow unbound variables to be used in expressions at all, and variables can only be used once they have been assigned, or bound to, a value. This use is also important in REDUCE, and it is achieved as in other languages by using the assignment operator, which in REDUCE is ":=". This is the same symbol as used in the ALGOLs, Pascal and Modula, whereas in FORTRAN, C and most BASICs it is just "=". A variable can be unbound or made free again by clearing it using the CLEAR command, as in the following example:

```
Y;

y

Y := A*X**2 + B*X + C;

          2
y := a*x   + b*x + c
```

```
Y;

   2
a*x  + b*x + c

CLEAR Y;  Y;

y
```

Notice that an assignment statement terminated by ";" is output as a full assignment statement, unlike all other statements that just output their values. An assignment statement has a value, which is the value of its right side, and this value can be used, such as by assigning it to another variable, so that the following are allowed:

```
A := B := C := 1;

a := b := c := 1

Y := (X := D**2) + 1;

        2
y := d  + 1

X;

 2
d
```

In common non-functional languages (other than C) statements do not have values, whereas in REDUCE many statements have useful algebraic values (and all have Lisp values).

A novel feature of assignments in REDUCE, which novice programmers are often surprised to find does not work in conventional languages, is that they can be made retrospectively. This is a consequence of the fact that REDUCE variables can remain symbolic, and are re-evaluated *in the current environment* whenever they are used in expressions. For example:

```
Z := Y;

z := y

Y := X;

y := x

Z;

x
```

```
CLEAR Y;  Z;
```

```
y
```

```
CLEAR Z;  Z;
```

```
z
```

Another novel feature of assignments in REDUCE is that *expressions* can have values assigned to them, e.g.

```
COSEC(X) := 1/SIN(X)$
SIN(X)**2 + COS(X)**2 := 1$
```

This provides a shorthand for an operation that is slightly different from assignment to a simple variable; it sets up a simplification or global[16] substitution rule, and is identical to what is more usually accomplished with the LET statement to be introduced in the next chapter. However, LET statements have much more general forms that have no assignment analogues; the assignment forms shown above work *only* for the specified variable names used in their definitions so that, for example, COSEC(Y) still has no simplification defined.

An assignment statement is easily confused with an equation. The difference mathematically is that an assignment statement is imperative – give the left side the value of the right side – whereas an equation is declarative – it is true that the values of the left and right sides are the same. Moreover, an equation is symmetrical, whereas an assignment is not. REDUCE has a rudimentary notion of an equation[17] as an algebraic object, for which it uses the operator "=". The symmetry of this notation is a nice reminder of the symmetry of an equation, in contrast to the asymmetry of the notation ":=" and the meaning of assignment. However, there is not very much at present that REDUCE can do with equations except construct them and take them apart again, and their main use is as a way of packaging information. An equation essentially has the status of an expression, but the "=" operator has no simplification rules – it would be nice if one could perform arithmetic on equations such that any operation were automatically performed on both sides, but at present this is not available.

It is quite permissible to type

```
Y = X**2;
```

```
   2
y=x
```

[16]The meaning of *global* is explained in §3.7.4.
[17]New in REDUCE 3.3.

but if, in fact, this was intended to be an assignment then the response from REDUCE gives very little clue (other than the lack of a space either side of the "=") that it was mistyped, so beware! You can also do this:

```
Z := Y = X**2;

         2
z := y=x
```

after which Z can be used as a shorthand for the equation to which it evaluates.

The left and right sides of an equation can be extracted using the operators LHS and RHS, so that with the above assignment to Z we get:

```
LHS(Z);

y

RHS(Z);

 2
x
```

1.9 Polynomials and rational expressions

The basic type of an algebraic expression in REDUCE is a rational expression, i.e. a quotient of polynomials, which internally is called a *standard quotient* (see chapter 5); any expression can be regarded as a special case of a rational expression, and is regarded so by REDUCE. A polynomial is just a rational expression with denominator 1. The coefficients of the polynomials can be symbolic, and can themselves be polynomials in other variables, which is how REDUCE represents multivariate polynomials and rational expressions. The default choice of which variable appears at what level in this nested structure is system dependent. Frequently it does not matter, but sometimes it may have a significant effect on speed of computation, and there are commands (ORDER and KORDER)[18] for specifying the ordering to be used. Unfortunately, not much is known about what is the best order to use in a particular problem (Pearce and Hicks 1981).

There are slightly finer distinctions relating to the representation of numerical coefficients, which is the subject of the next section, but we defer further details of the internal representations in general until

[18]See the REDUCE manual (Hearn 1987). Rayna (1987) also discusses uses of these commands in some detail.

chapter 5. Polynomial coefficients can also be functions, with arbitrary arguments (more exactly, the functions are treated as additional variables in the polynomials: see §5.3). REDUCE will put the arguments[19] into its standard internal representation, and may occasionally change the form of a function slightly, as in the example in §1.6.4. Alternatively, the outermost object can be a function, so that the expression is some function of a rational expression, rather than a rational expression of some functions. Examples of these two alternatives are the following:

$$\frac{\mathrm{fn}(x)+\mathrm{fn}(y)}{\mathrm{fn}(x)-\mathrm{fn}(y)} \quad \text{and} \quad \mathrm{fn}\left(\frac{x+y}{x-y}\right).$$

There is an incestuous relationship (mutual recursion) between the representation of rational expressions and other functions, which allows arbitrarily complicated expressions to be constructed. But whereas REDUCE knows how to simplify a rational expression to a standard form, and by default does so, it knows relatively little about other functions, and nothing at all about user-defined functions other than what the user tells it. The general name used in REDUCE for the data type of algebraic expressions, of which variables and constants are just special cases, is *scalar*.

1.10 Number domains and switches

We have seen that numbers can be input to REDUCE in various formats: integer, rational and real (i.e. floating-point). What is important is what REDUCE then does with them when simplifying expressions, and unless an expression contains only numerical constants, which is not how algebra systems are usually used, it regards numbers other than integer exponents as polynomial coefficients. Integer and rational are the fundamental numerical representations, and for most purposes REDUCE will by default manipulate them in the expected way. The only exception to this is that, by default, it regards a polynomial with rational coefficients as a rational expression with integer coefficients, which therefore cannot always be manipulated as a polynomial.

The operation of REDUCE is controlled by the setting of *switches*, which are turned on and off by the commands `ON` and `OFF`. The advertised switches are listed at the back of the REDUCE manual (in Appendix D) together with their default settings, which are usually

[19]The output format for arguments is controlled by the switch `INTSTR` – see §2.10.2.

off. To cause REDUCE to accept a polynomial with rational coefficients, and not convert it into a rational expression, make the switch setting `ON RATIONAL`. The need for this will become clearer later when we look at how to dissect polynomials and rational expressions.

Integer arithmetic can be performed modulo a given base, by specifying the base (e.g. 7) and turning on modular arithmetic, thus:

```
SETMOD 7;   ON MODULAR;
```

The order of these two commands does not matter, and the modulus is not evaluated modulo any previously set modular base. `SETMOD` returns the previous base, and `SETMOD 0` returns the current base without changing it; however, the returned base is evaluated modulo the current base if `MODULAR` is on[20], which can be misleading! Note that modular arithmetic applies only to polynomial coefficient-type calculations and not to exponents, and division by a number not co-prime to the modulus will cause an error.

Usually results in computer algebra must be exact, so exact integer and rational number representations must be used. However, sometimes it is necessary to work with numbers that are unavoidably irrational, and these can necessarily only be approximated. (They can alternatively be manipulated symbolically, and we mention the REDUCE Algebraic Number package in chapter 8.) For example, one may want numerical values from some algebraic expression that has been computed exactly, or one may need to compare the values of two expressions in a way that does not require exact values but does require explicit numerical values. We introduced in §1.6.2 the two real (i.e. floating-point) number domains provided in REDUCE, namely float and bigfloat.

Float mode is selected by `ON FLOAT` and uses the standard floating-point facilities provided by the computer[21]; probably the same as FORTRAN single-precision, which will usually give 6 or 7 significant decimal digits of accuracy. Zero is always represented by integer 0. By default, floats that are equal to integers are converted[22] to integer representation, but this can be prevented by switching `OFF CONVERT`.

Bigfloat mode is much more powerful (but much slower) than float mode, and is selected by `ON BIGFLOAT`. This mode uses a precision of 10 significant decimal digits by default, which can be changed to (say) 15 by `PRECISION 15`. `PRECISION` is actually an operator that

---

[20] We have suggested a correction for this – see the Appendix, §G.

[21] There are bugs in float mode in Atari REDUCE – see the Appendix, §A.2.

[22] See the Appendix, §A.3, for details.

returns as its value the new precision[23], except in the special case of `PRECISION 0`, which returns the current precision without changing it.

It should be noted that "`PRECISION` n" does not mean that calculations will be performed using precisely n significant digits, but rather that REDUCE will try to output results *accurate* to n significant digits. In order to achieve this, more digits are used internally (usually 2 more than the specified precision), and changing the precision will not immediately change the number of digits that are stored internally. (There is no easy way to force REDUCE to perform arithmetic to precisely n significant figures, as might be desirable to investigate problems in numerical analysis, although it is possible to simulate this behaviour fairly well by interacting with the bigfloat package at a low level using symbolic mode.) Note also that the floating-point input routine currently accepts input in float format only, which is then converted into bigfloat mode as appropriate, but with limited precision; this will be corrected in future versions.

In rational, float and bigfloat modes REDUCE regards the denominator of a number as the integer 1, so that the value resides entirely in the numerator, which in general will not be an integer, in contrast to the default mode in which a number is represented as an integer numerator and a positive integer denominator. The above number domains are mutually exclusive; when a new number domain is turned on any number domain that was in effect is automatically turned off. All numbers should be automatically converted into the current representation, although at present bigfloat numbers do not seem to be converted automatically. The conversion of floating-point numbers to rational representation may not always produce quite the expected result; the details of the conversion between float and rational (or integer) representation are discussed in the Appendix, §A.3.

The next two modes[24] can be used (in principle) to qualify any of the above domains. By default, REDUCE regards a complex number, specified in the usual way, e.g. `A + I*B`, as a polynomial in the variable `I` with a built-in rule that `I**2` simplifies to -1. To tell REDUCE that the fundamental number domain is complex, for purposes such as factorization, switch `ON COMPLEX`. This causes complex numbers to be represented in a special form, which depends also on the current number domain, and not as polynomials in `I`.

To force complex numbers to have their denominators rationalized to real form (which is independent of the `COMPLEX` switch) and rational

---

[23] It might be more useful for it to return the old precision – see §§4.14, 4.15

[24] New in REDUCE 3.3.

powers in denominators to be rationalized, switch `ON RATIONALIZE`. For example, this switch setting causes automatic rationalization of `1/(A+I*B)` to `(A-I*B)/(A**2 + B**2)` and `5/SQRT(5)` to `SQRT(5)`.

1.11 Predefined variables and functions

The variables `T` and `NIL` are best avoided in algebraic code because they represent respectively true and false in Lisp, as will be discussed further in chapter 5. The variables `E`, `I` and `PI` have useful predefined algebraic properties, and are intended to represent the exponential number, the base of imaginary numbers and the circular constant respectively; they are best avoided for most other purposes. They are most important in conjunction with the predefined functions, which are

```
             sqrt exp log
   sin cos tan cot asin acos atan
  sinh cosh tanh asinh acosh atanh
           dilog erf expint.
```

For example, REDUCE knows that

```
exp(1) = e,  e**(i*pi) = -1,  log(1) = 0,  log(e) = 1,
        sin(0) = 0,  sin(pi) = 0,  sin(pi/2) = 1,
e**x = exp(x),  sin(-x) = -sin(x),  x**(1/2) = sqrt(x),
```

and can also differentiate and integrate these functions. It knows a few other properties at this level of sophistication, but the best way to find out whether REDUCE knows some particular property of these functions is to evaluate an expression that uses the property and examine the result. We will show later how to define properties of functions in REDUCE. One reason why REDUCE does not know more such properties is that there is no guarantee that particular replacements will be required in all circumstances; e.g. it is probably useful to simplify $\sin\frac{1}{4}\pi$ to $1/\sqrt{2}$, so this rule is built into REDUCE, but it may not always be desirable to simplify $\sin\frac{1}{8}\pi$ to its rather messy expression in terms of square roots; therefore that simplification is not built in.

There is often not much that can usefully be done to simplify transcendental functions unless their arguments are numerical, in which case the functions can be explicitly evaluated, but in general this unavoidably implies using numerical approximations. REDUCE will numerically evaluate non-arithmetic functions only if the switch `NUMVAL` is turned on[25] *and a floating-point number domain has been selected*; it will then numerically approximate at least the following functions

```
sin cos tan asin acos atan sqrt exp log **
```

and the predefined symbolic constants `E` and `PI`. (REDUCE itself supports numerical evaluation of these functions and symbolic constants in both float and bigfloat modes independently of the underlying Lisp, although in some implementations some of these evaluations may be passed down to the Lisp system.)

For example:

```
ON NUMVAL, BIGFLOAT;
SQRT 2;

1.414 21356 2

PI;

3.141 59265 4

E;

2.718 28182 8
```

As this example shows, REDUCE breaks bigfloat numbers into groups of five characters from the left for output. (It might be better if it grouped them away from the decimal point.)

Float mode should be significantly faster than bigfloat mode, especially on a computer that has hardware floating-point support, and float mode is recommended whenever its range and accuracy suffice.

REDUCE also has built in the function **ABS** which returns the absolute value of its argument if it evaluates to a number, but remains symbolic otherwise[26]. The "numerical" functions **MIN** and **MAX** applied to an arbitrary number of integer arguments[27] return respectively their minimum and maximum values, as in FORTRAN.

---

[25]In REDUCE 3.3 it is off by default, but we understand that in future versions it will be on by default.

[26]New in REDUCE 3.3, and contrary to what the manual currently states, although **ABS** applied to a non-numerical rational expression is currently not simplified correctly, and when applied to a complex number it does not currently evaluate to the modulus, both of which problems are easily fixed. In previous versions **ABS** simply evaluated to its argument if it was non-numerical.

[27]Code can easily be developed to support **MIN** and **MAX** with arbitrary arguments – see later exercises and the Appendix, §G.

1.12 Lists

All the more structured objects available in REDUCE are built from scalar data types. The most fundamental structured type is the *list.* This is fundamental in that it mimics the Lisp list that is the fundamental data structure on which the whole of REDUCE is built (cf. chapters 5 and 6), and it is also the fundamental way of packaging information to be passed to and from several major REDUCE facilities. Algebraic-mode lists are a major new feature of REDUCE 3.3; they provide a much more elegant mechanism for some of the tasks for which global arrays, matrices or variables had to be used in previous versions.

A list is built by putting the elements of the list, separated by commas, inside braces (curly brackets) thus:

```
{1, 0, X, 1, X**2, 2};

            2
{1,0,x,1,x ,2}
```

The elements of a list can be almost any REDUCE objects including lists, e.g.

```
{{1, 0}, {X, 1}, {X**2, 2}};

                  2
{{1,0},{x,1},{x ,2}}
```

Lists can be nested within lists to arbitrary depth. The objects placed inside a list will be evaluated before they are stored, and so can be bound variables or expressions that will be replaced by their values, as well as unbound variables or constants. Thus there is no need for a special operator (like the `LIST` function in Lisp) to build lists from objects that must be first evaluated – this is done automatically by the code generated by the `{ ... }` construct. This is the main external difference between REDUCE algebraic-mode lists and Lisp lists.

It is often necessary to build a new list out of an old one other than by simply including the old one inside the new one as we have done above. The fundamental way of extending a list is to add another element to the front of it. This is fundamental because it is simplest in view of the way that a list is stored internally as a chain of pointers. This method of building a list is called *constructing* it, and the operator that does it is therefore called `CONS`, following a long-established Lisp tradition. In Lisp terminology (cf. chapters 5 and 6) the head of a list and the whole of its tail are sometimes called a *dotted pair* because of the way they are usually written

```
( head . tail )
```

The operation of "consing" an element onto the front of a list constructs a dotted pair, so the CONS operator is represented in REDUCE[28] as the *infix* operator ".", which in use looks exactly as above. It is safest to put a space either side of the ".", especially if numbers are involved, when the "." might be misinterpreted as a decimal point, as illustrated below. The empty or null list is denoted by {} in REDUCE (but by () or NIL in Lisp). Hence one could construct a list of numbers like this:

```
LIST := {};

list := {}

LIST := 0 . LIST;

list := {0}

LIST := 1 . LIST;

list := {1,0}

LIST := 2 . LIST;

list := {2,1,0}

% But beware of omitting spaces:
LIST := 3.LIST;

*** 3.0 represented by 3/1

list := 3*{2,1,0}
```

The above example is a silly way to construct a list that could have been written directly, but it illustrates a technique that is very powerful when used together with the iterative control structures that will be introduced later, for example, to construct a list of objects that satisfy a recurrence relation.

Note that CONS (".") sticks a new element onto the front of a list; there is no *efficient* way to put a new element anywhere else in a list, although the operator APPEND joins two lists together, thus:

```
APPEND({0, 1, 2}, {3, 4, 5});

{0,1,2,3,4,5}
```

---

[28]However, REDUCE algebraic mode does not support dotted pairs that are not also lists, whereas Lisp does.

This provides a way of adding an element to the end of a list:

```
APPEND({0, 1, 2}, {3});

{0,1,2,3}
```

but note that the element to be added to the end must first be made into a list, and that `APPEND` has to do a lot more work than `CONS`.

We have now seen all three of the basic ways of building a list: writing it in braces "`{}`" explicitly, adding elements to the front using `CONS` ("`.`") and joining lists using `APPEND`. Elements also need to be extracted from lists, and the fundamental way to do this uses the operator `FIRST`, which returns the first element. This is fundamental in the same way that `CONS` is. The operator `REST` returns what is left of a list after the first element has been removed, which is *always* a list, although it may be the empty list. The operators `FIRST`, `REST` and `CONS` are related by

list = FIRST(*list*) . REST(*list*)

There are also operators `SECOND` and `THIRD` that return the second and third elements of a list, and could be expressed as

SECOND(*list*) = FIRST(REST(*list*))
THIRD(*list*) = FIRST(REST(REST(*list*)))

For example:

```
LIST := {1, 2, 3, 4, 5}$
FIRST(LIST);

1

REST(LIST);

{2,3,4,5}

SECOND(LIST);

2

THIRD(LIST);

3
```

The more general operator PART(*list*, n) returns the nth element of *list* (if it exists), where n must evaluate to an integer. The number of elements in a list is returned by LENGTH(*list*), which does *not* recursively count the elements of any lists that are elements of the outermost list. Finally, the operator `REVERSE` reverses the elements in a list, but again *not* those within any lists that are elements of the outermost list. For example:

```
LIST := {1, 2, 3, {4, 5}}$

PART(LIST, 4);

{4,5}

LENGTH(LIST);

4

REVERSE(LIST);

{{4,5},3,2,1}
```

The operator PART is a convenient but inefficient and inelegant way of accessing the elements of a list; it is reasonable to use it interactively but best avoided within program loops. However, it does allow elements to be easily accessed from the *end* of a list by using a negative part argument, and it allows an element in a list to be changed in a way that would be awkward to achieve in any other way. This usage is a little unusual; the changed list is returned as the value of an assignment with a PART operator on the left, like this:

```
LIST1 := {A, B, C}$
LIST2 := PART(LIST1, 2) := Z;

list2 := {a,z,c}
```

PART can also be used with multiple part arguments to pick out part of a nested structure – see the REDUCE manual (Hearn 1987), §§8.5.3, 8.5.4.

In REDUCE 3.3 there is a restriction on assignments involving lists: if a list has been assigned to a variable then that variable must be cleared before any other type of data can be assigned to it, and vice versa. We understand that the type-checking mechanism will be changed in future versions of REDUCE to remove this unexpected over-protectiveness of lists.

1.13 Solving equations

We finish this chapter with two major REDUCE facilities that make essential use of expressions, equations and lists. The REDUCE operator SOLVE will attempt to solve either a single equation in one unknown, or a coupled set of linear equations in several unknowns. The interface that SOLVE uses would handle the general case of coupled nonlinear equations, and this observation helps one remember how the interface works, but the solution routines cannot at present handle the general case. In chapter 8 we will look at a package that can be used for solving coupled polynomial equations, although similar facilities for more general problems such as solving coupled transcendental equations are still under development (e.g. Davies and Wright 1991).

In fact, SOLVE can either be given an equation A = B to solve, or it can be given the equivalent expression A-B and will try to find its zeros. So when we refer to an "equation" we will mean either of these two forms of input. The syntax of SOLVE is one of

SOLVE(*Equation*, *Variable*)
SOLVE(*EquationList*, *VariableList*)

if there is respectively one or more equations. If the variables are unambiguous, i.e. all constants are explicit rather than symbolic, then *Variable* or *VariableList* may be omitted and SOLVE will deduce for itself what the unknowns are. Hence the following are all equivalent:

```
SOLVE(X**2 = 1, X);

{x=-1,x=1}

SOLVE(X**2 = 1);

unknown: x

{x=-1,x=1}

SOLVE(X**2 - 1);

unknown: x

{x=-1,x=1}
```

but this fails:

```
SOLVE(X**2 = A);

unknowns: {x,a}

***** non linear equation solving not yet implemented
```

The operator SOLVE always returns any solutions that it finds as a list[29], which is the *value* of the SOLVE operator. It returns actual solutions in the form of equations with the unknowns isolated on the

[29]In previous versions it returned the *number* of solutions found, and put them into a global matrix called SOLN.

left, as shown above; if it cannot solve a particular factor then it returns that factor equated to zero, thus:

```
SOLVE(X**7 - X**6 + X**2 = 1, X);

  6
{x  + x + 1=0,x=1}
```

In the case of a single equation SOLVE returns a list of equations representing the solutions, but for a coupled set of equations it returns a list of lists, each of which represents one solution, e.g.

```
SOLVE({X + 3*Y = 7, Y - X = 1}, {X, Y});

{{x=1,y=2}}
```

The philosophy is that SOLVE returns a list of solutions, where a solution in 1 variable is represented as a single equation with the variable on the left, whereas a solution in n variables is represented as a *list* of n equations, each with one of the variables on the left, so that each individual *solution* has the same form as the first argument of SOLVE. This mechanism would be necessary when returning the solutions of a coupled nonlinear system, which would in general have multiple multivariate solutions. (For coupled *linear* equations there would be no ambiguity in returning a simple list of equations, as for a single nonlinear equation, but with *different* unknowns on the left of each one. However, the more general mechanism has been chosen, presumably with a view to future enhancement of SOLVE.)

Roots of a nonlinear equation (or system) may be repeated or multiple, and SOLVE assigns to the global variable MULTIPLICITIES!* a list[30] of the multiplicities in 1-to-1 correspondence with the roots, e.g.

```
SOLVE( (X-1) * (X-2)**2 * (X-3)**3 );

unknown: x

{x=3,x=2,x=1}

MULTIPLICITIES!*;

{3,2,1}
```

The number of distinct solutions can be found by applying LENGTH to the list of solutions returned by SOLVE.

You may by now be wondering how SOLVE works. For a single equation it recursively uses square-free factorization (see chapter 7) together with the known inverses of log, sin, cos, **, acos and asin, and

[30] A column matrix in previous versions.

the formulae for the zeros of quadratic, cubic and quartic factors (which become progressively more tortuous and therefore less useful). For coupled linear equations **SOLVE** merely provides an alternative interface to the built-in matrix facilities, which we will consider in their own right in the next chapter.

1.13.1 Non-unique solutions

A consistent singular equation such as $x = x$, or system such as $x + y = 0$, $2x + 2y = 0$, has a non-unique solution involving one or more parameters, e.g. $x = \alpha$ or $x = \beta$, $y = -\beta$ respectively for the above two examples, where α and β may be complex. By default, REDUCE introduces arbitrary variables (more precisely, operator references: see §2.11) called **ARBCOMPLEX**(j), where j is an explicit integer, as it needs them in such solutions. Within any REDUCE session the index j starts at 1, and is automatically increased by 1 every time that REDUCE needs a new arbitrary parameter[31]. Introduction of these parameters can be prevented by setting **OFF SOLVESINGULAR**.

Similarly, equations involving functions with multi-valued inverses generate solutions with arbitrary real or integer parameters; these are called **ARBREAL**(j) and **ARBINT**(j) respectively by REDUCE and are handled the same[32] as **ARBCOMPLEX**(j). To prevent the introduction of such parameters, and thereby select only the principal branches of the inverse functions, set **OFF ALLBRANCH**.

1.14 Factorizing

Factorizing expressions is closely related to solving equations. By default, REDUCE expands all expressions by multiplying out products and powers. Alternatively, if **ON FACTOR** is set REDUCE will attempt to factorize all expressions. Factorization is attempted over the currently selected number domain if it is appropriate, which includes complex (i.e. Gaussian integers) and modular domains but excludes the floating-point domains. (Polynomials over the rationals are the same as over the integers, apart from an overall constant.) Factorization is fully multivariate, so there is no need (and hence no facility) to specify variables with respect to which to factorize, and symbolic constants are regarded as variables. However, at present REDUCE will not factorize multivariate polynomials over a modular domain. Note that a

[31] If necessary, j can be reset, its value being that of the symbolic-mode variable `!!ARBINT`.

[32] They share the single sequence of argument values j.

side-effect of setting ON FACTOR is that REDUCE turns off the switch EXP, which may have unexpected consequences – see §2.10.1.

If it is necessary to manipulate the factors of an expression further then it is more convenient to put them into a list, which is what the operator FACTORIZE(*expression*) does[33]. The value of this operator is a list of polynomials which are the irreducible factors of *expression* (i.e. they cannot be further factorized over the current number domain). The number of factors found can be determined by applying LENGTH to the list of factors.

However, there is an inconsistency in the output from FACTORIZE. If there is an overall numerical factor (other than 1), the so-called *numeric content* of the polynomial, then that will appear as the first element of the list, and by default it will not itself be factorized further, e.g.

```
FACTORIZE(X**2 - Y**2);

{x - y,x + y}

FACTORIZE(4*X**2 - 4);

{4,x - 1,x + 1}
```

Therefore, before processing the list of factors it is necessary to decide what the first element is. When working interactively with REDUCE one can just look at the list and decide; when calling FACTORIZE inside a larger piece of code (a REDUCE "program") a test such as NUMBERP can be used, as we will see later. This must be done before the result of applying LENGTH to the list can be interpreted.

In order to have the numeric content factored set ON IFACTOR, after which the prime factors of the content will occupy as many as necessary of the initial elements of the list returned by FACTORIZE, e.g.

```
ON IFACTOR;  FACTORIZE(4*X**2 - 4);

{2,2,x - 1,x + 1}
```

(This switch does not affect factorized output produced by setting ON FACTOR.) Numbers on their own (as special cases of polynomials) can be factorized if the IFACTOR switch is on. However, applying FACTORIZE to either 1 or 0 produces an empty list regardless of the setting of IFACTOR – for 1 this is a special case of a numeric content of 1 not producing an explicit factor.

Another point to watch when using FACTORIZE, especially in pro-

[33] In previous versions it used an array or a set of scalars and returned the number of factors as its value.

grams, is that the order and signs of the polynomial factors in the list are system dependent, and so cannot be relied upon, although any non-trivial numeric content is always the first element, and the product of the signs of the factors is correct. The factorizer works by first reducing multivariate problems to univariate ones and then solving them modulo small primes that are generated randomly – see chapter 7. (Use of a particular prime can be forced by giving it as a second argument to FACTORIZE.)

1.15 Exercises

The only way to learn a practical subject such as computer programming is to do it, and to try things that you think should work. If they do, that's fine; if they don't then you can learn a lot by finding out why not. The exercises at the end of each chapter are intended to be done using only facilities introduced in that and preceding chapters, and you should try to use only those facilities, although it may be possible to do them more elegantly or more easily using facilities that will be introduced later. The earlier exercises are intended to be solved in a naïve interactive way – don't waste too much time trying to produce very elegant and sophisticated "programs". As an example of using only elementary facilities even though more sophisticated facilities exist, it is possible to evaluate an expression EXPX in X at X = A using only assignments as follows:

```
X := A$ EXPA := EXPX; CLEAR X$
```

[Comments like this in square brackets are "asides" that may be hints or remarks about practicalities that are not directly part of the question.]

1. Try starting and stopping REDUCE, and check whether *Control-c* or some other control character will interrupt it. Try entering a few single-letter variable names in upper and lower case, terminated by either ; or $ and observe the differences in output.

2. See how deeply you can nest an exponentiation of the form x^{x^x} and still get REDUCE to differentiate it and integrate it back again in a reasonable length of time. [This will depend on the amount of main memory available.] You can generate suitable integrands by using the recurrence relation

$$y_{n+1} = x^{y_n}, \text{ where } y_1 = x.$$

3. Find the minimum point and value of $f(x) = x^2 + 4x + 5$.

4. Calculate the coefficient of x^3 in the Taylor series expansion about 0 of $\exp(x + x^2)$. [This does not require the rest of the Taylor series.]

5. Find dy/dx when

$$(a)\ y = \frac{1+\sin x}{1+\cos x}, \qquad (b)\ y = \ln\sqrt{\frac{1+x}{1-x}},$$

$$(c)\ y = \sec\cos x, \qquad (d)\ y = \ln\left(\frac{(1+x)^{1/2}}{(1-x)^{2/3}}\right), \qquad (e)\ y = \sin|x|$$

and try to simplify the results as far as possible. Remember that in REDUCE you can assign a value to an *expression* in order to force simplification. We will return to the problem posed by the last expression in §3.6 – it is intended only to point out a problem at this stage!

6. Find

$$(a)\ \int \cos^3 x\,dx, \qquad (b)\ \int x^2 e^x\,dx, \qquad (c)\ \int \frac{dx}{(x-1)(2x+1)},$$

$$(d)\ \int \frac{7}{(3-x)(1+2x)}\,dx, \qquad (e)\ \int_0^1 \sqrt{4-x^2}\,dx.$$

If REDUCE fails on the last integral, try `ON ALGINT`, and if it still fails [meaning that the ALGINT package is not loaded – see §8.2], or as an alternative approach, use the substitution $x = 2\sin\theta$.

7. Find the solution of the differential equation $e^{2x+y}dy/dx = x$ that satisfies $y = 0$ at $x = 0$. [This has to be done "by hand", but using REDUCE instead of pen and paper – there is no built-in differential equation solver, but see §8.6.]

8. A curve is given parametrically by $x = \sin t$, $y = \cos^3 t$, for $-\pi < t \leq \pi$. Find the gradient of the curve at $x = 0$ and find all values of t at which the curve has inflexions (i.e. vanishing second derivative).

9. Experiment with long and/or obscure identifiers. Can you find a limit on the length of identifiers? Are there any non-alphanumeric characters that can be included without using the escape character "`!`"?

10. Experiment with numbers – try entering various integer, rational and floating-point numbers, and see what REDUCE does with them. Can you find the largest integer and floating-point numbers that can be entered from the keyboard? Can you generate larger numbers, such as by raising a large number to a large power? How does REDUCE output string constants? Can they be assigned to variables? Can they be used in any other way, e.g. as symbolic constants?

11. Is `E**X` the same as `EXP(X)`? Is `X**(1/2)` the same as `SQRT(X)`? Try these for some numerical values of `X` with `ON NUMVAL` and a suitable number domain. Is `ON FLOAT` sufficient, or do you need `ON BIGFLOAT`?

12. Does the statement `X = Y` change the value of the variable `X`? How about `X := Y`? Beware – the difference is easily overlooked! Convince yourself that a multiple assignment of the form `A := B := C := 1` really assigns the same value to all the variables.

13. Find the values of π, e, $\sqrt{2}$, $\sqrt[3]{2} \equiv 2^{1/3}$, $\log 2$, and any other values that take your fancy, to 100 significant figures. Check the values by applying a suitable "inverse" function to the results, e.g. apply a few trigonometric functions to π, take the log of e, etc.

14. Given that $z_1 = 1 + i\sqrt{3}$ and $z_2 = 4 + 3i$, find the moduli and arguments of $z_1 z_2$ and z_2/z_1.

15. Construct a list of 5 elements each of which is a list whose single element is the previous element, and whose first element is 0, i.e. `{0, {0}, {{0}}, ... }`. Can you find a neat recursive algorithm for doing this? Apply `LENGTH` and `REVERSE` to the result – are the results of doing this what you expected?

16. If the roots of $x^2 + 5x + 3 = 0$ are α and β, construct an equation with roots α^2 and β^2 and check that it is correct.

17. Show that the equation $e^x \cos 2x = 1$ has a root between 0.4 and 0.45. See if REDUCE can solve the equation. If not, use the Newton-Raphson algorithm interactively to improve the accuracy of the root. (See also §4.14.)

18. Find the general solution of the equation $7 \sin x + 24 \cos x = 15$. [You may have to apply a trigonometric sum formula to give REDUCE some help!]

19. By applying half-angle formulae to $\sin\frac{1}{4}\pi$ etc. construct exact expressions that are equivalent to $\sin\frac{1}{8}\pi$ etc. but involve only fractional powers of integers.

20. If $z = \ln(x^2 + y^2)$ prove that $(\partial z/\partial x)^2 + (\partial z/\partial y)^2 = 4e^{-z}$.

21. If $y = x^n \ln x$ prove that $x\, dy/dx = x^n + ny$.

22. Construct explicitly the differential equation

$$(1-x^2)\frac{d^2y}{dx^2} - 4x\frac{dy}{dx} - (1+x^2)y = 0$$

and then show that $y = (\sin x)/(1-x^2)$ is a solution.

23. By differentiating implicitly and then solving for the derivative, find dy/dx from each of the following equations:

$$(a)\ x^2+y^2-4x+3y+2=0, \qquad (b)\ \ln(x^2+y^2) = 2\arctan\left(\frac{y}{x}\right).$$

24. Factorize $x^n - 1$ for various n.

25. Marin Mersenne (who lived from 1588 to 1648) asserted that the only primes p for which $M_p \equiv 2^p - 1$ is prime are 2, 3, 5, 7, 13, 17, 19, 31, 67, 127 and 257. It is now known that M_{67} and M_{257} are not prime, but M_{61}, M_{89} and M_{107} are prime. Verify some of these assertions for *small p* – for large p you may have a *very* long wait!

2

Controlling REDUCE; operators and matrices

We begin with some technicalities of the interactive control of REDUCE, then consider how to manipulate polynomials and rational functions in more detail, how to perform substitutions and control simplification, and how to control the internal and external forms of expressions. Finally, we introduce two more structured data types: operators and matrices.

2.1 Screen control

REDUCE regards a terminal screen as a window onto an infinitely long roll of paper, and does not have any concept of the depth of the screen. Therefore, it will happily scroll output without stopping for you to read it, and there is no facility within *standard* REDUCE to control output paging. This is left to the computer operating system and/or underlying native Lisp. Some systems (e.g. 386-REDUCE) provide an automatic paging system, but most do not. On some systems it is possible to stop the output by typing *Control-s*, and restart it by typing *Control-q* (or sometimes any key). If this procedure works then it should merely pause the output to give you time to read it, without actually losing any, but especially when using a terminal connected via a long network link strange things can happen. It may sometimes be necessary to hold down *Control-s* briefly until it has been noticed – this is the case on the Atari.

2.2 Interactive workspace access and editing

When using REDUCE interactively it is often useful to be able to refer back to what you did previously, and REDUCE provides several mechanisms for doing this. To look at previous interactive input, REDUCE provides the DISPLAY command: DISPLAY followed by an explicit positive integer constant n will display the last n statements or commands[1], whereas DISPLAY ALL will display all of them (scrolling them rapidly up the screen!).

The algebraic result of any previous statement can be accessed using WS, which is an abbreviation of *WorkSpace*. Used alone as a variable WS evaluates to the last algebraic result. When followed by a number n it evaluates to the algebraic result of the nth input statement, if that statement produced one. REDUCE stores all final algebraic results, and WS simply recalls the appropriate stored value and does not re-execute the statement that produced it.

The result of a previous statement can be assigned to a variable for easier access, or to an expression to establish a global substitution rule, either by using a normal assignment statement with WS or WS n on the right, or in the case of the last algebraic result the SAVEAS command can be used as an alternative to assigning WS. The effects of the following two statements are identical:

variable_or_expression := WS
SAVEAS *variable_or_expression*

except that SAVEAS does not return any algebraic value. SAVEAS can also be used in a more general form – see §2.8.2.

The nth input statement or command can be re-executed by giving the command "INPUT n". This is useful for re-executing commands and statements that do not produce immediate algebraic results. If it is used to re-execute a statement that does produce an algebraic result then that result will be recomputed, and may be different from the result produced previously. If this is not the intention then it is more efficient to use "WS n". Note that both WS and INPUT can be used within expressions, and have the status either of a variable in the case of WS alone, or of functions if they are followed by a number (cf. §2.12).

Often it would be nice to be able to re-execute a previous statement or command but with a small change (particularly after having mistyped something!). REDUCE contains, in most implementations, a rudimentary input editor for this purpose, which is invoked by the

[1] In input order in REDUCE 3.3, but in reverse order in previous versions. An explicit positive float argument will be treated like the smallest integer not less than it; any other argument is equivalent to ALL. Better use could be made of this argument flexibility – see the Appendix, §G.

command ED[2]. If ED alone is typed as a command then it attempts to edit the previous input, which is particularly useful for correcting typing mistakes; if "ED n" is typed, where n is a positive integer, then it attempts to edit the nth input statement or command.

The standard editor is very crude, partly to make it portable. Some of the least portable features of computers are the details of keyboard and screen handling, and therefore the standard editor is not a display or screen editor, but is statement oriented and only semi-interactive in that it does not immediately display the result of any edit command. (Note that some computers, terminals, or programming environments provide facilities for recalling and editing previous input, which may be much easier to use than ED, and are worth finding out about!)

When invoked, ED displays the input that it has been asked to edit, followed by a ">" prompt for edit commands. ED operates on an internal buffer, and has a pointer to the current character in this buffer. All commands operate from the current character to the end of the statement, and a command consists of a single character. A string of commands can be given in one line, but nothing happens until the Return key has been pressed. There are commands to insert or search for a string of characters terminated by an "escape" character, which is usually generated by the *Escape* key[3] and is *not* the same escape character ("!") as used to escape special characters in identifiers.

The set of commands available can be displayed at any time by giving the help command "?" (followed by *Return*), and the best way to learn to use ED is to practice with it. The commands are fairly mnemonic, but for reference they are reproduced in Table 2.1 as they would be displayed on screen (although note that they should work equally well if typed in either case). The command E means execute the edited statement, which automatically re-displays it first, and the command Q means quit or abandon the edit, which is useful if you have accidentally started editing the wrong input, or completely messed it up and want to start again.

Note that in Table 2.1 the notation "**<something>**" is a description of what should be typed: "**<character>**" means any single character; "**<string>**" means a sequence of one or more characters etc. – the "<" and ">" are not part of the command.

The command "EDITDEF *ProcedureName*" invokes ED on the named

---

[2]It is possible to redefine ED to invoke a different editor, which will be highly system dependent. We assume here that the standard input editor – called internally CEDIT – is what ED invokes.

[3]But not always, because at least one operating system treats *Escape* specially for its own purposes; try *Control-g* as a second guess.

Table 2.1. Help output from the standard input editor ED

```
THE FOLLOWING COMMANDS ARE SUPPORTED:
   B               move pointer to beginning
   C<character>    replace next character by <character>
   D               delete next character
   E               end editing and reread text
   F<character>    move pointer to next occurrence of <character>
   I<string><escape>    insert <string> in front of pointer
   K<character>    delete all chars until <character>
   P               print string from current pointer
   Q               give up with error exit
   S<string><escape> search for first occurrence of <string>
                       positioning pointer just before it
   <space> or X    move pointer right one character

ALL COMMAND SEQUENCES SHOULD BE FOLLOWED BY A CARRIAGE RETURN
    TO BECOME EFFECTIVE
```

procedure[4]. However, the procedure is displayed in a form much closer to its internal Lisp representation than to what the user actually typed, particularly for algebraic procedures, and EDITDEF cannot be used at all on compiled procedures (unless the source form is kept, which in standard REDUCE is not done). Moreover, it is not possible to save the result of the editing in any current version of EDITDEF that we have tried, making it of use only for displaying a procedure.

2.3 File input and output

Non-trivial use of REDUCE almost certainly needs input from and output to files, although this will still usually be controlled interactively. Filenames must be either identifiers or strings in double quotes. If any non-alphanumeric characters are required in a filename, such as directory separators, drive identifiers etc. it is usually easier to use a string, e.g. "G:\FILES\INPUT.RED", rather than to precede all the special characters with "!", e.g. G!:!\FILES!\INPUT!.RED. This example would suit an MS-DOS or Atari system with a hard disc, but the structure of filenames is highly system dependent. Some systems *require* REDUCE files to have a particular extension (such as the .RED above), and may even automatically add it (whether or not you have already specified it, leading to a possible source of errors), but most allow *any* valid filename. Filenames are handled by the system-dependent part of

[4]For a description of procedures, see chapter 4.

REDUCE, and the degree of sophistication is left to the implementer[5].

REDUCE input can be taken from a file instead of from the keyboard by giving the command "IN *filename*". If the IN command is terminated with ";" then the contents of the file will be displayed as it is read; if the terminator is "$" it will not. (This display or "echoing" can also be controlled by using the switch ECHO *within the file being input*[6], which over-rides the terminator used for the IN statement and applies only while that file is input.) Files to be INput to REDUCE should contain exactly what would have been typed at the keyboard, except that it is recommended that they end with ";END;". But beware that "END" serves rather a lot of purposes in REDUCE, and if REDUCE is not expecting an END then it will cause a switch into Standard (almost the native) Lisp mode (which is the level below symbolic mode: see chapter 6).

Input files can themselves contain IN commands, so that hierarchical structures of input files can be built up. At the end of reading an input file REDUCE will switch back to the previous source of input so that input files can be nested. If the reserved name T is used as a filename then input is taken from the terminal until "END;" is explicitly typed, which could be useful for providing temporary keyboard input while primarily inputting from a file, but see also the discussion of XREAD in §3.8.

The command PAUSE in an input file (only) will suspend input from the file, switch input back to the terminal and display the message "CONT? (Y OR N)". The response Y (or y) will continue inputting from the file, whereas N (or n) will leave it suspended and take further input from the keyboard. This behaviour is also provoked by some errors in input files[7]. At any future time the command CONT will continue input from the suspended file. This provides both a way of simply pausing input from a file so as to read it or its results, and also the advertised way of switching temporarily back to keyboard input *in the middle* of a file. The equivalent of a FORTRAN 77 PAUSE, e.g. for use within a procedure or loop, can be simulated using XREAD[8].

---

[5]But, like everything else in REDUCE, file handling can be customized by the enthusiastic user – see the Appendix, §G.

[6]ECHO is normally off – turning it on for terminal input has rather strange effects, and is not recommended!

[7]This can be avoided by setting the switch ON ERRCONT, which is particularly useful if REDERR (see §4.2) is being explicitly called by the user from an input file as a way of handling error conditions.

[8]See §3.8. This is analogous to the recommended technique in FORTRAN anyway, since PAUSE is considered "obsolescent" in FOR-

Like most REDUCE commands, IN can be given a sequence of filenames separated by commas, which REDUCE will read in the specified order before finally returning to the keyboard. The arguments to IN are not evaluated and so must be explicit filenames. One thing that may be surprising about using IN interactively is that only the IN command itself is stored; neither the contents of the input file nor the results it produces are stored and so these are not accessible by DISPLAY, ED, INPUT or WS.

Output can be directed to a file with the command "OUT *filename*", a special case of which is "OUT T" which directs output to the terminal. When output to a file is completely finished the file should be closed with the command "SHUT *filename*"; if this is not done then some output may be lost, although on normal termination REDUCE will close all open files. Output can be switched among files and the terminal without shutting any of the files, and when output is switched to an open file it is added to the end of that file (i.e. it is appended). However, if output is switched to a file that has been shut, or which has not been opened in the current REDUCE session, then the file is truncated to zero length thereby *destroying* any information that was previously in it before the output is added (i.e. the file is over-written).

At present there is, unfortunately, no facility within standard REDUCE to explicitly direct output to more than one place at once, so it is not possible to keep a record in a file of the output appearing on the terminal screen. (An exception is that warning and most error messages are also copied to the terminal when output is directed to a file – see the end of §2.5.) However, some operating systems (e.g. some versions of UNIX) do provide such a "transcript" facility which can be used with REDUCE, and some implementations (notably Acorn- and 386-REDUCE) support simultaneous output to both the screen and a file.

When output is switched with OUT then whatever would have appeared on the screen appears instead in the selected file, and can be controlled by the choice of terminator just as when outputting to the screen. If it is intended to read information back in that has been output to a file, then the output has to be in the correct format with expressions in "linear" rather than "two-dimensional" style. The switch setting OFF NAT switches off the "natural" or two-dimensional output usually used, and produces output that can be re-input[9].

TRAN 90.

[9]With the exception that in the current version the escape character "!" is omitted, and would need to be edited back in by hand where necessary. We have suggested a correction for this.

It is often convenient to develop REDUCE programs interactively and then save the *input* to a file for later use, probably after editing. The only standard way[10] to do this is

```
OUT FileName;
DISPLAY ALL;
SHUT FileName;
```

and all the prompts must then be edited out of the result before it can be re-input.

Perhaps the best way of keeping a record of the solution of a problem using REDUCE is as follows. Develop the necessary statements and commands interactively, save them to a file and edit them as necessary. Then, assuming that the final "perfect" REDUCE program is in the file *InFile*, a complete record of its operation can be produced in the file *OutFile* by

```
OUT OutFile;
IN InFile;
SHUT OutFile;
```

The file *OutFile* will now contain a "perfect" transcript of the REDUCE solution of the problem except that there will be no prompts, making it a little difficult to distinguish output from input. One resolution of this difficulty is always to type input in the opposite case (see §1.4) from that in which REDUCE produces its output. This is how we produced most of the example sessions in this book. An alternative would be to use "batch" mode – see §3.10.

2.4 Online lessons and manual access

REDUCE is supplied with a lot of documentation files, but whether they are available online, and if so their location and precise names, will be system dependent. There is a set of 7 "online lesson" files (by D. Stoutemyer) that are intended to be input into REDUCE using the `IN` command. They give a brief description of a topic followed by an invitation to try some real examples. They contain some excellent advice, but are hard to read on many systems because they are not "paginated" and scroll through too quickly. You may find it easier to read them using a file editor or file perusal program.

The complete REDUCE manual is supplied as a file, and if a suit-

---

[10]But see the Appendix, §G.

able program is available it can be conveniently read online[11]. A keyword search can help to find topics of interest quickly. REDUCE also comes with its complete source code, most of which is written in RLISP. This is the only source of documentation about the internal operation of REDUCE (as far as we are aware) apart from this book, the later interactive lesson files, and a few articles that sketch the general ideas. Finally, there are example and test files for most of the major facilities in REDUCE.

2.5 Error messages and recovery

REDUCE error messages are preceded by "*****", and usually cause termination of the current calculation and a return to a new prompt.

```
% Demonstration of error messages:
ON IGLOO;

***** igloo not defined as switch

DEPEND Y, 1;

***** 1 invalid as kernel

DEPOND Y, X;

***** syntax error: depond(y),x invalid
```

After a non-syntactic error, such as attempting some operation in an inappropriate number domain, it may be possible to correct the error and then repeat the statement that failed by giving the command RETRY:

```
% Demonstration of "RETRY":
DEG(X**2/2, X);

        2
       x
***** ---- invalid as polynomial
       2

ON RATIONAL;  RETRY;

2
```

Warning messages are preceded by "***" and either do not stop execution or, in interactive mode, REDUCE may ask the user for clarification before continuing.

[11]But see the Appendix, §B.

```
ARRAY A(1);
ARRAY A(2);

*** array a redefined

UNDECLARED!-OPERATOR(X);

*** undeclared-operator declared operator

undeclared-operator(x)
```

When REDUCE discovers a syntactic error in an input statement, i.e. something that it simply does not understand, it indicates where it thinks the error is by reprinting the statement with the error marked by the symbol "$$$". Interactive input can be corrected by invoking ED, but if inputting from a file it will be necessary to edit the file and re-input it. However, for immediate use, if the error causes a pause as described in §2.3, the erroneous command could be re-typed at the keyboard followed by the command CONT. In principle, it is always possible to continue the input process, but in our experience this is not usually helpful because one error, especially in a procedure, can cause many spurious subsequent errors.

Unfortunately, syntax errors are frequently not well localized by the parser, as the following examples show:

```
IF Y1 = Y2 THIN 1 ELSE 0;

else$$$ 0;

***** else invalid in if statement
```

where in fact THIN should have been THEN.

```
FOR ALL THETA LOT COS THETA ** 2 + SIN THETA ** 2 = 1;

for all theta lot cos theta**2+sin theta**2=1$$$;

***** ; invalid in let statement
```

where in fact LOT should have been LET.

If output is redirected to a file then a second copy of error and warning messages is also sent to the terminal (or standard output stream), with one exception in REDUCE 3.3. Messages about type errors in the form

```
***** X invalid as Y
```

are sent only to the current output stream, which we suspect is due to an oversight that could easily be corrected. [To be specific, there are two versions of the internal procedure TYPERR responsible for outputting this message – the version defined in the RLISP source file behaves like the other message procedures, whereas the version defined in the ALG1 source file outputs only to the current stream. The latter outputs algebraic values in algebraic rather than internal format. Which version is used is likely to depend on which REDUCE modules have been loaded.]

2.6 Dissecting polynomials and rational functions

The general data structure in REDUCE is the rational expression, which is a quotient of two polynomials. A rational number is just a special case of this. The operators **NUM** and **DEN** return respectively the numerator and denominator of a rational expression; applied to an expression that is not rational they return respectively the expression itself and 1. The precise effect of these operators for numbers depends on the number domain in force, but whatever it is the numerical factor returned by **DEN** is *always* an integer, thus:

```
R := 22/7$
NUM(R);

22

DEN(R);

7

ON RATIONAL;
NUM(R);

 22
----
 7

DEN(R);

1

ON FLOAT;

*** domain mode rational changed to float

NUM(R);

3.142857

DEN(R);

1
```

Polynomials in REDUCE can be multivariate, but a multivariate polynomial is just regarded as a univariate polynomial in one of the variables with coefficients that are polynomials in the remaining variables. REDUCE actually stores multivariate polynomials internally in this "recursive" representation (see §5.3 for further details), which allows information about a multivariate polynomial to be extracted using functions defined for univariate polynomials, by specifying the variable with respect to which the polynomial is to be regarded as univariate. Only the REDUCE factorizer works directly on multivariate polynomials (at least as it appears to the user); all other operators for decomposing polynomials treat them as univariate, and therefore generally need to know which variable is being regarded as the unknown.

Perhaps the most obvious information to extract from a polynomial is a list of all of its coefficients, and the operator `COEFF(`*poly*`, `*var*`)` does precisely that, returning as its value a list[12] of coefficients ordered by *ascending degree*, e.g.

```
P := A*S**3 + B*S**2 + C*S + D$
COEFF(P, S);
```

```
{d,c,b,a}
```

As a side effect, `COEFF` also sets two useful global variables to the highest power (degree) and the lowest power (order), thus:

```
HIPOW!*;
```

```
3
```

```
LOWPOW!*;
```

```
0
```

To find just the coefficient of a particular power of a variable the operator `COEFFN(`*poly*`, `*var*`, `*power*`)`[13] is provided, e.g.

```
COEFFN(P, S, 2);
```

```
B
```

`COEFF` is almost certainly more efficient if you actually want several coefficients. These two operators are more general than the other polynomial operators in REDUCE and will accept coefficients with *symbolic*

[12]New in REDUCE 3.3 – in previous versions `COEFF` used a global array or constructed identifiers for the coefficients and returned the degree as its value.

[13]New in REDUCE 3.3.

denominators[14] as long as the denominator does not contain the unknown (this restriction is removed if the switch **RATARG** is turned on – see below).

Four other polynomial decomposition operators relate more closely to the internal representation (see §5.3) than does **COEFF** and all operate with respect to the variable *var* given as the second argument:

- **DEG**(*poly*, *var*) returns the degree;
- **LCOF**(*poly*, *var*) returns the leading coefficient;
- **LTERM**(*poly*, *var*) returns the leading term;
- **REDUCT**(*poly*, *var*) returns the reductum.

These operators should satisfy the following identities:

$$\mathrm{lterm}(poly, var) \equiv \mathrm{lcof}(poly, var) * var^{\mathrm{deg}(poly, var)},$$

$$poly \equiv \mathrm{lterm}(poly, var) + \mathrm{reduct}(poly, var),$$

which they do when *poly* really is a polynomial in *var*, but in the special case that *poly* is a constant – either an explicit numerical constant or a symbolic constant independent of *var* – then they do not. This is due to minor inconsistencies in the code[15] that are very easy to correct, although the definition of **LTERM** given in the manual is also inconsistent with that of **LCOF**, the latter being the only operator that is completely correct.

As an example of using these operators, suppose that we have two polynomials p and q regarded as univariate polynomials in x and we want to eliminate the leading terms between the two polynomials. The result could be found using the following code:

```
d := deg(p,x) - deg(q,x);
result := lcof(q,x)*p*x**max(0,-d) -
          lcof(p,x)*q*x**max(0,+d);
```

But since this expression has been designed so that the leading terms cancel there is little point in explicitly subtracting them as above. More particularly, if for some reason one is working over a floating point number domain the cancellation may not occur exactly. Therefore, for both the above reasons, the following code is probably better:

---

[14] New in REDUCE 3.3.

[15] We understand that these inconsistencies will be corrected in future versions – see also the Appendix, §D.3 and §G.

```
result := lcof(q,x)*reduct(p,x)*x**max(0,-d) -
          lcof(p,x)*reduct(q,x)*x**max(0,+d);
```

If the switch RATARG[16] is turned on then the above 6 operators will accept arbitrary rational expressions and treat their denominators as symbolic constants, *even* if the denominators depend on the polynomial unknown. This is currently documented only for COEFF and COEFFN.

Some related built-in operators on pairs of polynomials are the following, of which the first two also work for integers:

- GCD(*poly1*, *poly2*) returns their greatest common divisor;
- REMAINDER(*poly1*, *poly2*) returns the remainder when *poly1* is divided by *poly2*;
- RESULTANT(*poly1*, *poly2*, *var*) returns the result of eliminating *var* from the equation pair *poly1* = *poly2* = 0 (or 0 if the two polynomials have a non-trivial gcd).

The operator PART provides access to the values of parts of expressions, and can also be used to change them. It is not easy to use, except in the particularly useful case of lists as discussed in §1.12, because it accesses the parts of an expression in terms of a combination of the internal representation and the positions in which the parts are (or would be) output. The latter depends on various switch settings, the most important of which are described in §2.10. Whilst this facility may be convenient interactively, it is unlikely to be reliable in general REDUCE procedures or "programs". It is described in detail by Rayna (1987), §2.11 – see also the REDUCE manual (Hearn 1987), §§8.5.3, 8.5.4.

In a similar spirit, the operator TERMS applied to an algebraic expression returns the number of terms in the numerator of the expression (subject to the current switch settings EXP and MCD – see §2.10.1). The operator LENGTH applied to an algebraic expression returns the number of additive top-level terms in its expanded representation, which in this case includes the denominator. For example:

```
Z := (A*B + C*D)/(A*C + B*D);

       a*b + c*d
z := -----------
       a*c + b*d

TERMS Z;     =>  2
LENGTH Z;    =>  4
```

[16]New in REDUCE 3.3.

2.7 Local substitution (SUB and WHERE)

Substitution covers several operations that might not be thought of mathematically as substitution. As in conventional mathematical notation, a function $f(x)$ defined as a REDUCE operator or procedure can be evaluated at any point $x = a$ by simply writing $f(a)$, and two functions can be composed by writing $f(g(y))$. If f is an *expression* rather than explicitly a function, then these operations involve a *substitution* for one of the variables in the expression.

The REDUCE operator

SUB(*var1* = *exp1*, *var2* = *exp2*, ..., *expression*)

evaluates to *expression* with *var1* replaced by the expression *exp1*, etc. Arguments to REDUCE operators are nearly always fully evaluated before an operation is applied to them, and *all* the objects in the argument list of SUB, including the equations, can be variables that have had appropriate values assigned to them. However, each *var* argument must finally evaluate to an unbound identifier or "indecomposable expression"[17]; it is not possible to substitute for *general expressions* (sums, products etc.) using the SUB operator (it requires a let-rule, as explained in the next section, or possibly an assignment) and it is not possible to substitute for explicit constants. Here are some simple examples:

```
P := Z**2 + Z + 1;

       2
p := z  + z + 1

% Evaluate p at z=0 and assign the result to p0:
P0 := SUB(Z=0, P);

p0 := 1

% Evaluate p at z=x**2; equivalently compose p with z(x):
P1 := SUB(Z=X**2, P);

       4    2
p1 := x  + x  + 1

% Just to show that EVERYTHING gets evaluated first:
S := SOLVE(X**2 = 1);

unknown: x
```

[17]Formally, a *kernel*, which is explained at the end of §2.9 and in §5.3.

```
s := {x=-1,x=1}

SUB(FIRST(S), X**2 - 1);

0

SUB(SECOND(S), X**2 - 1);

0
```

A single SUB will make all the substitutions "in parallel", i.e. all in the *original* expression rather than in partially substituted forms of it. To force sequential substitution use nested calls of SUB; note the difference between the following two statements:

```
SUB(X=Y, Y=X, X/Y);

 y
---
 x

SUB(X=Y, SUB(Y=X, X/Y));

1
```

A very useful feature of SUB that is currently undocumented is that the equations can be in one or more (possibly nested) lists, such as those returned by SOLVE; we will illustrate the use of this in §4.13. However, the final expression cannot be a list or any other structured object at present.

A different kind of local substitution is provided by the WHERE[18] operator, which has the syntax

expression WHERE *var1* = *exp1*, *var2* = *exp2*, ...

To some extent this is like an infix version of

SUB(*var1* = *exp1*, *var2* = *exp2*, ..., *expression*)

but WHERE differs from SUB in that only the *expi* on the right of the equations are evaluated before making the substitutions, so that WHERE only substitutes for *explicit* occurrences of the *vari* in *expression*, which is itself only evaluated *after* the substitutions have been performed. The *vari* are effectively local to the WHERE operator.

The WHERE construct is intended as a mechanism for writing an expression in terms of named common sub-expressions, such as

```
(X + Z)/(X - Z) WHERE Z = INT(LOG(X), X);
```

[18]New in REDUCE 3.3.

```
      log(x)
 - ------------
   log(x) - 2
```

This particular example could just as well be written as

```
SUB(Z = INT(LOG(X), X), (X + Z)/(X - Z));
```

which gives exactly the same result as long as Z is unbound; otherwise the SUB version is not reliable. However, an example that *cannot* be written using SUB, because it requires the substitution to be performed before the expression is evaluated, is this (the facilities used in this and the next example will be explained in §§3.4 and 2.8 respectively):

```
ON NUMVAL, FLOAT;

(IF W > 0 THEN SQRT(W) ELSE -(-W)**(1/3))
   WHERE W = SIN(23);

 -0.9458621
```

These WHERE substitutions can all be viewed another way: the expression can be regarded as defining an anonymous function of the WHERE-variables, and the WHERE applies this anonymous function to the arguments that appear on the right of the equations. Thus the previous example may be regarded as a local and anonymous version of the following:

```
OPERATOR ANON;
ON NUMVAL, FLOAT;

FORALL W LET ANON(W) =
     IF W > 0 THEN SQRT(W) ELSE -(-W)**(1/3);
ANON(SIN(23));

 -0.9458621
```

Such anonymous functions are quite commonly used in Lisp programming, where they are called *lambda expressions*, and they are supported in RLISP – see §6.8. The WHERE operator provides a lambda construct in algebraic mode, which can also be used as a synonym for a lambda expression in symbolic mode (whereas SUB is not defined in symbolic mode).

2.8 Global substitution and simplification rules (LET)

When manipulating trigonometric formulae one would, almost without thinking, replace any occurrence of the combination

$$\cos^2\theta + \sin^2\theta$$

by 1. This is a *global* substitution or simplification that could also be effected by replacing $\cos^2\theta$ by $1-\sin^2\theta$ (or vice versa). But if you saw the combination

$$\cos^2\theta - \sin^2\theta$$

would you make the same replacement? Depending on the context, it might be best to replace this combination by any one of the equivalent formulae

$$1 - 2\sin^2\theta,\quad 2\cos^2\theta - 1,\quad \cos 2\theta.$$

This example shows that there are really very few global substitutions that can be guaranteed to be the right ones to make in all circumstances, and therefore REDUCE has very few of them built in. It is left to the user to set up the appropriate rules, and it is usually necessary to clear them again quite soon. They are established with a LET statement, and are therefore often referred to as "let-rules". For example:

```
Y := COS(X)**2 + SIN(X)**2;

            2          2
y := cos(x)  + sin(x)

LET COS(X)**2 + SIN(X)**2 = 1;  Y;

1

CLEAR COS(X)**2 + SIN(X)**2;  Y;

       2          2
cos(x)  + sin(x)
```

The general syntax of a simple let-rule is

LET *lhs1* = *rhs1*, *lhs2* = *rhs2*, ...

Although the rules look symmetrical they are not; the left side is always replaced by the right side. The rules are cleared by listing *only* the left sides as arguments to the CLEAR command, thus:

CLEAR *lhs1*, *lhs2*, ...

A new let-rule with the same left side as an existing rule will probably replace it in effect, but it is safer to clear the old rule first. It is often best to establish a let-rule, use it, and then clear it immediately, as illustrated in §2.9; this avoids the possibility of establishing ambiguous combinations of let-rules, the order of application of which is not reliable.

We have seen that let-rules can, and usually do, have ***general expressions*** on their left sides which are to be replaced, something that is not allowed when making local substitutions with **SUB**. However, internally the replacement rules are modified so that there is just one "irreducible" term to be replaced. This means that

```
LET COS(X)**2 + SIN(X)**2 = 1;
```

will be represented internally as either $\cos^2 x \to 1 - \sin^2 x$ or $\sin^2 x \to 1 - \cos^2 x$, but which is used is system dependent[19], so if it matters then the correct rule should be specified explicitly. It also means that reciprocals are not allowed on the left side, e.g. `LET 1/Y = X` fails because REDUCE attempts to represent it internally as $1 \to xy$ which is clearly ambiguous, whilst the equivalent rule `LET Y = 1/X` works and so does `LET Z/Y = X`.

One other small catch to be aware of is that let-rules for **EXP(X)** and **SQRT(X)** must be established in terms of **E**X** and **X**(1/2)**, otherwise they will not work (cf. exercise 6 of chapter 5).

Let-rules and assignment statements appear to have a similar syntax and to do similar things, and it is very important to be clear about the distinction between them. An assignment "*lhs* := *rhs*" assigns the *current value* of *rhs* to *lhs*, thereby changing the value of *lhs*. Introducing a let-rule by "LET *lhs* = *rhs*" does nothing in itself, in the sense that the *rhs* is *not* evaluated and if the rule were cleared again immediately nothing would have changed. A let-rule only has an effect when an expression containing *lhs* is evaluated: then it causes *lhs* to be replaced by the current value of *rhs*. The value used is the value of *rhs* at the time the let-rule is used, not the value at the time it is set. The replacement of *lhs* is done repeatedly until *lhs* no longer appears, so unqualified self-referential let-rules will cause infinite loops in the simplifier, whereas self-referential assignments work in REDUCE just as in any other language. Some comparative examples should make all this clearer:

```
% Assignment:            Let-rule:
% ----------             --------
X := A$                  XX := AA$
X := X + B$  X;          LET XX = XX + BB;  XX;

a + b                    ***** simplification recursion
                               too deep


A := 1$                  AA := 1$
```

[19]This is discussed in more detail by Rayna (1987), p. 84.

```
D := (A+B)**2$  D;       LET DD = (AA+BB)**2;  DD;

 2                         2
b  + 2*b + 1             bb  + 2*bb + 1


A := 2$  D;              AA := 2$  DD;

 2                         2
b  + 2*b + 1             bb  + 4*bb + 4


CLEAR A;  D;             CLEAR AA;  DD;

 2                         2                 2
b  + 2*b + 1             aa  + 2*aa*bb + bb
```

If the replacement is only to be made when *lhs* stands as a separate term (i.e. not multiplied by other factors) then in place of LET one can use MATCH. This works just like LET except that the term replaced must exactly match *lhs*. Match-rules are cleared like let-rules.

Assignments in REDUCE are slightly different from those in other languages in that they can be applied retrospectively and expressions can appear on the left – they are really "immediate" versions of the simple let-rules introduced above. However, the following generalized forms of let-rule have no analogous assignment form.

2.8.1 Using let-rules to establish identities

What distinguishes an identity from an equation is that the former is true for all values of its variables; the variables are just place-holders, and their particular names are irrelevant. REDUCE can be told that the names of variables used in let-rules are irrelevant by using the FOR ALL (or equivalently FORALL) construct, thus:

FOR ALL *var1*, *var2*, ... LET *lhs1* = *rhs1*, *lhs2* = *rhs2*, ...

For example,

```
FOR ALL THETA LET COS(THETA)**2 + SIN(THETA)**2 = 1;
FOR ALL A, B LET LOG(A*B) = LOG(A) + LOG(B);
```

It is also possible to establish restricted identities using qualified let-rules of the form

FOR ALL *varlist* SUCH THAT *condition* LET ...

to which we will return in §3.6 when we have considered conditional expressions. All let-rules are cleared by repeating the let-rule with LET replaced by CLEAR and all components of the form "= *rhs*" omitted, thus:

```
FOR ALL THETA CLEAR COS(THETA)**2 + SIN(THETA)**2;
FOR ALL A, B CLEAR LOG(A*B);
```

and for the qualified form

FOR ALL *varlist* SUCH THAT *condition* CLEAR ...

However, there are a couple of provisos. Firstly, a let-rule of the form

```
FOR ALL N LET X**N = ...
```

will never match either x ($\equiv x^1$) or 1 ($\equiv x^0$), which means that in some situations a lot of special-case rules must be established, as illustrated by the example at the end of the next section (§2.9).

Secondly, there are some global simplifications that cannot be specified using let-rules, for example:

```
FOR ALL A,B LET LOG(A) + LOG(B) = LOG(A*B)

***** unmatched free variable(s) =b
```

This arises because, as we mentioned earlier, all but one of the terms in a sum on the left of a let-rule are taken over to the right. In this case, LOG(B) is taken over to the right, leaving no term in B on the left. This problem may happen with other similar kinds of sum simplifications that the current REDUCE pattern matcher cannot handle, and how to do this generally is still a research problem. It is not hard to write a (symbolic) procedure to perform such a simplification locally, but this is more in the spirit of Maple and is left as an exercise in symbolic-mode programming for the reader (see chapter 9). Alternatively, there is an experimental implementation of a more powerful pattern matcher (based on that in SMP, which uses pattern matching more extensively than does REDUCE) available in the REDUCE Network Library[20], which *might* help with this kind of problem.

2.8.2 Using WS and SAVEAS to establish identities

When using REDUCE interactively, it can be very useful to be able to convert the result of a previous calculation into a substitution rule without having to retype it. Therefore, a special case of a let-rule in which the right side *is* evaluated at the time that the rule is defined is when the right side is WS, which allows the last algebraic result to be made the result of a substitution rule. However, note that this works

[20]See the Appendix, §F.

only with `WS` used as a variable; the form `WS` n used on the right of a let-rule causes an internal error (at least in Atari REDUCE). Moreover, just as an assignment of `WS` can be effected by the `SAVEAS` command (see §2.2), so can a let-rule with `WS` as its right side, in either the simple or completely general form. Thus the effects of the following two statements are identical:

```
FOR ALL varlist SUCH THAT condition LET var_or_exprn = WS
FOR ALL varlist SUCH THAT condition SAVEAS var_or_exprn
```

and both forms are cleared by

```
FOR ALL varlist SUCH THAT condition CLEAR var_or_exprn
```

A trick for forcing evaluation of the left side of a let-rule is introduced in §2.13 at the end of this chapter, and explained in a little more detail in §3.7.1.

2.8.3 Resubstitution and complete evaluation

REDUCE will continue to apply substitution rules to an expression until no more match, regardless of whether the substitution rules were established using assignment, `LET`, `MATCH` or `SAVEAS`. Equivalently, it will continue to try to re-evaluate a variable until the resulting expression evaluates to itself. Usually this is desirable and what one would expect, but just occasionally it would be nice to have some control over the depth of evaluation or substitution, e.g. to be able to go one step at a time, rather than all the way in one go. Unfortunately there is currently no simple reliable way to do this.

Versions of REDUCE prior to 3.3 advertised a switch called `RESUBS`, which was on by default, and if turned off would inhibit all "resubstitutions". However, the REDUCE manual made it clear that "because of the complexity of the substitution mechanisms" the effect of turning off `RESUBS` was not entirely reliable. For that reason, `RESUBS` has not been declared to be a switch in REDUCE 3.3, and is not mentioned in the REDUCE 3.3 manual. Nevertheless, it still exists internally, and it can still be turned off and on using the `OFF` and `ON` commands. They will produce an error message, but they will still change the value of the internal logical switch variable (which is generally true for switches that exist internally but are not declared to be switches). Rayna (1987, §3.5) gives an example of the use of `OFF RESUBS` (in REDUCE 3.2), but even that very simple example does not work correctly in REDUCE 3.3.

2.9 Controlling polynomial degree

When a polynomial in the variable *var* is simplified subject to a let-rule of the form "LET *var***pow* = ..." every occurrence of *var***pow* as a *factor* in each term is replaced by the right side of the rule. This can be used (with care) to control polynomial degree. For example, when developing power series expansions one usually wants terms only up to some fixed maximum degree, and computing unwanted higher-degree terms would be inefficient. When working with univariate power series approximations, a let-rule of the form "LET *var***pow* = 0" will cause all powers of *var* equal to *pow and higher* to be set equal to 0, and so discarded, e.g.

```
LET X**3 = 0;   (X+1)**5;

    2
10*x  + 5*x + 1
```

But note that such simplification rules are applied at all intermediate stages in a computation. Compare the following:

```
LET X**8 = 0;   X**10 / X**5;

0
% instead of x**5 as expected.

X**20 / X**10;
***** zero denominator
% instead of 0 as expected.

% To get the expected results requires ...
CLEAR X**8;
Q1 := X**10 / X**5;

         5
q1 := x

Q2 := X**20 / X**10;

         10
q2 := x

LET X**8 = 0;
Q1;

 5
x

Q2;

0
```

It is often the case that let-rules must be set only when required and cleared as soon as they have been used, otherwise they can produce effects that are neither expected nor desired! A fairly common idiom is therefore this:

```
P := (X+1)**5$
LET X**3 = 0;
P := P$
CLEAR X**3;

P;

     2
10*x  + 5*x + 1
```

The apparently redundant assignment "P := P" in the new environment (after the let-rule) is necessary to make the effect of the rule on the value of P permanent after the rule has been cleared.

If it is necessary to replace specific powers of a variable only, i.e. to regard monomials[21] as irreducible objects, rather than products of lower-degree monomials, this can be achieved by the use of MATCH, e.g.

```
% Regard w as the principal 8th root of unity
P := (1+W)**8;

       8       7        6        5        4        3        2
P := w  + 8*w  + 28*w  + 56*w  + 70*w  + 56*w  + 28*w

      + 8*w + 1

MATCH W**2 = I, W**4 = -1, W**6 = -I,
      W**5 = W**(-3), W**7 = W**(-1);

ON DIV;  P := P;

                   -3    8        -3
P := - (48*i*w    - w  - 48*w    + 69)
```

As we have seen, REDUCE is perfectly capable of handling multivariate functions, and one might well want to work with a multivariate power series approximation such as a multivariate Taylor polynomial. There are several ways of setting the maximum degree in this case. The statements

WEIGHT *var1* = *wt1*, *var2* = *wt2*, ... ;   WTLEVEL *d*;

set a maximum weighted degree of d. This means that a monomial of the form $var1^{p1} * var2^{p2} * \cdots$ is set to 0 if $p1 * wt1 + p2 * wt2 + \cdots > d$.

---

[21] A monomial is one term of a polynomial (sometimes taken to exclude the coefficient; see §8.3.1).

The default weight level corresponds to setting **WTLEVEL 1** (not 2 as stated in the manual), which retains only linear terms. This weight mechanism works fine in simple situations, but it has the disadvantage that, because of the way the facility is implemented, variables that have weights set cease to be kernels. The weights are removed by

```
CLEAR var1, var2, ... ;
```

A kernel in REDUCE is any object that can be treated as the unknown in a polynomial, i.e. that cannot be considered as a sum or product. For example, any global identifier that has nothing assigned to it is a kernel, as is an operator of any form that remains unevaluated, such as COS(X**3 + A + 1) or DF(Y,X) when Y has been declared to DEPEND on X. But X**3 + A + 1 is not itself a kernel. The WEIGHT statement shown above replaces the variable *var1* internally by the *term var1*(K!*)**wt1*, which is not a kernel, although the output routines never output the factor involving K!*. It is now fairly obvious how the facility works: WTLEVEL D is equivalent to LET (K!*)**(D+1) = 0. Only kernels can appear on the left of the equations in SUB statements, and REDUCE will only SOLVE for kernels or differentiate or integrate with respect to kernels. Hence it may be necessary to set weights and clear them again repeatedly.

An alternative is to implement the same facility explicitly, along the following lines, assuming that expression depends on var1 and var2:

```
expression := sub(var1 = var1*wght**wt1,
                  var2 = var2*wght**wt2, expression)$
let wght**(d+1) = 0;
% Perform the calculation involving expression
clear wght**(d+1);
expression := sub(wght = 1, expression)$
```

It is still necessary to keep very careful track of the weightings, and performing complicated calculations with multivariate degree cutoff is not trivial!

Another method of implementing multivariate degree cutoff is even more explicit. Here is an example, in which a cutoff above degree 3 is imposed uniformly, i.e. without weighting, on the two variables X and Y:

```
% General case:
FOR ALL P1, P2 SUCH THAT P1 + P2 > 3 LET
   X**P1 * Y**P2 = 0;

% Special cases:
LET X**3 * Y = 0, X * Y**3 = 0;
LET X**4 = 0, Y**4 = 0;
```

```
% Try it:
Z := (1+X+Y)**4;

         3         2          2          2
z := 4*x  + 12*x *y + 6*x  + 12*x*y  + 12*x*y + 4*

            3      2
     x + 4*y  + 6*y  + 4*y + 1

% And to prove that x is still a kernel:
SUB(X=0, Z);

   3      2
4*y  + 6*y  + 4*y + 1
```

Although the condition that we have used in the let-rule above will not be formally introduced until the next chapter, its meaning should be fairly obvious.

This technique has the advantage that there is no need to manipulate a subsidiary "weight-carrying" variable when working with X and Y, and indeed this degree cutoff mechanism should be completely transparent (apart from potential problems with intermediate expressions as illustrated earlier for the univariate case). However, it has the disadvantage that many more let-rules must be established, simply because the general monomial in a let-rule will not match all cases, as mentioned in the previous section. The technique merits further consideration, but not here. It is easy to add non-uniform weighting if required.

2.10 Controlling expression representation

Changing the internal representation of expressions will tend to change the output format, but not vice versa. Some control over both is provided by switches, most of which control output format only. The more important ones are presented below, in terms of the effect of changing them from their default settings. Attempting to set a non-existent switch is detected as an error.

2.10.1 Controlling internal representation

OFF EXP prevents the expansion of powers and products. This may, for example, facilitate the cancellation of common polynomial factors in a fraction. The disadvantage is that the representation is not *canonical* although it is still *normal* (see §§5.1 and 5.4), which means that two mathematically equal polynomials may not be identical in REDUCE, although their difference will still be zero:

```
% Demonstration of OFF EXP:
OFF EXP;  Y1 := (X+1)**2;

                 2
y1 := (x + 1)

ON EXP;  Y2 := Y1;

          2
y2 := x  + 2*x + 1

OFF EXP;  Y1;

         2
(x + 1)

% With OFF EXP zero expressions simplify to 0:
Y1 - Y2;

0

% but equal expressions do not compare equal:
LET SAME = (IF Y1 = Y2 THEN EQUAL ELSE UNEQUAL);  SAME;

unequal

ON EXP;  SAME;

equal
```

Note that a side-effect of setting `ON FACTOR` is that REDUCE turns off the switch `EXP` (because obviously the two have opposite effects). This means that with `ON FACTOR` set expressions may not be fully simplified, and can retain a "partially factored" form – see Rayna (1987), p. 140, but remember that not all of Rayna's remarks apply to REDUCE 3.3.

`OFF MCD` prevents REDUCE from putting a sum of rational expressions over a common denominator (`MCD` stands for "Make Common Denominator"), causing the representation not even to be normal:

```
% Demonstation of OFF MCD:
OFF MCD;  Y1 := 1/X - 1/(X+1);

                       -1    -1
y1 :=  - ((x + 1)   - x   )

ON MCD;   Y2 := Y1;

               1
y2 := -----------
        x*(x + 1)
```

```
% With OFF MCD zero expressions might NOT simplify to 0:
OFF MCD;  Y1 - Y2;

                 -1            -1
 - (x*((x + 1)  *x + (x + 1)    - 1))/(x*(x + 1))

ON MCD;   Y1 - Y2;

0
```

OFF MCD also causes effects similar to setting ON RATIONAL and ON DIV (see below), but these other switches do not have the potentially unpleasant consequences of OFF MCD, and should therefore be used rather than OFF MCD except in special circumstances.

ON GCD causes REDUCE to cancel all common polynomial factors in rational expressions, whereas by default only common numerical factors, variables and other "obvious" polynomial factors are cancelled (see Rayna (1987), §3.2, for details), and if EXP is also OFF then REDUCE attempts to factorize polynomials. However, in general this is only partially successful, and for complete cancellation or factorization ON FACTOR (which automatically switches OFF EXP) should be used instead of ON GCD. Beware that computing gcds or factorizing can be *very* time-consuming, especially if done at every step, as happens if the switches GCD or FACTOR respectively are turned on.

ON EZGCD causes an improved modular gcd algorithm to be used when GCD is ON instead of the default trial-division algorithm; the EZGCD algorithm (see chapter 7) is almost always significantly faster, but the code to support it is large, and it only works for integer coefficients. Alternatively, ON HEUGCD causes a heuristic gcd algorithm to be used.

By default, a gcd calculation is performed to ensure that when rational expressions are combined over a common denominator the least common multiple of the denominators is used, even if GCD is OFF. OFF LCM (LCM stands for "Least Common Multiple") prevents this.

REDUCE's treatment of non-integer powers is not entirely consistent. The expression X**(1/2) is treated in the same way as other non-integer powers for input, but is always output in terms of the SQRT function, whereas SQRT X is treated slightly differently for input. According to the REDUCE manual:

> With the default system switch settings, the argument of a square root is first simplified, and any divisors of the expression that are perfect squares taken outside the square root argument. The remaining expression is left under the square root. However, if the switch REDUCED is on, multiplicative factors in the argument of the square root are also separated, becoming individual

square roots. The switch REDUCED also applies to other rational powers in addition to square roots.

What REDUCE actually does is illustrated in Table 2.2, which is the slightly edited output from a short program included for reference in the Appendix, §B. This table shows that the REDUCED switch affects

Table 2.2. Effect of the REDUCED switch

```
                 6                               6
        x = 8*a *b*c                    x =  - 8*a *b*c
        ============                    ===============

                       3                                     3
sqrt x:  2*sqrt(2*b*c)*a                2*sqrt( - 2*b*c)*a

                                    3                        3
x**1/2:  2*sqrt(c)*sqrt(b)*sqrt(2)*a    2*sqrt( - 2*b*c)*a

          q  q  6*q  q                       6      q
x**q  :  c *b *a   *8                   ( - 8*a *b*c)

            1/3  1/3  2                          1/3  2
x**1/3:  2*c    *b   *a                 - 2*(b*c)    *a

            2/3  2/3  4                    2/3  2/3  4
x**2/3:  4*c    *b   *a                 4*c    *b   *a

ON REDUCED

sqrt x:  2*sqrt(c)*sqrt(b)*             2*sqrt(c)*sqrt(b)*

                       3                            3
             sqrt(2)*a                     sqrt(2)*a *i

x**1/2:  2*sqrt(c)*sqrt(b)*             2*sqrt(c)*sqrt(b)*

                       3                            3
             sqrt(2)*a                     sqrt(2)*a *i

          q  q  6*q  q                       6      q
x**q  :  c *b *a   *8                   ( - 8*a *b*c)

            1/3  1/3  2                     1/3  1/3  2
x**1/3:  2*c    *b   *a                 - 2*c    *b   *a

            2/3  2/3  4                    2/3  2/3  4
x**2/3:  4*c    *b   *a                 4*c    *b   *a
```

only square roots that are input explicitly as square roots, as described in the manual, and explicit fractional powers with *odd numerators* of explicitly *negative expressions*; all other cases are unaffected by the `REDUCED` switch, and except for symbolic powers are fully "reduced" by default. (This deduction appears to be borne out by inspection of the source code. However, algebraic numbers and functions are a difficult computational domain; we believe that this part of the simplifier is fairly heuristic but we admit that we do not properly understand it.)

`ON PRECISE` causes a square root such as `SQRT(A**2)` to be simplified to `ABS(A)`, whereas by default it would be simplified to `A`, so that `ON PRECISE` should be set if it is important to preserve the principal (positive) branch of the double-valued square-root function even if `A` is subsequently assigned a negative value. This switch applies *only* to square roots.

2.10.2 Controlling output format

We will introduce the main commands and switches through the following examples, which we hope will be clearer than trying to describe their effects:

```
% Construct a typical truncated power series:
% ===========================================
EX := FOR I := 1:5 SUM X**I/I;

                  4        3        2
        x*(12*x  + 15*x  + 20*x  + 30*x + 60)
ex := -----------------------------------------
                          60

% Try to make it look better:
% ---------------------------
OFF ALLFAC;  EX;

     5        4        3        2
 12*x  + 15*x  + 20*x  + 30*x  + 60*x
----------------------------------------
                   60

ON DIV;  EX;

 1   5    1   4    1   3    1   2
---*x  + ---*x  + ---*x  + ---*x  + x
 5        4        3        2

% One of the following two forms is probably optimal:
ON REVPRI;  EX;
```

```
     1  2    1  3    1  4    1  5
x + ---*x  + ---*x  + ---*x  + ---*x
     2        3        4        5

OFF RATPRI;  EX;

          2        3        4        5
x + 1/2*x  + 1/3*x  + 1/4*x  + 1/5*x

ON ALLFAC, RATPRI$  OFF DIV, REVPRI$

% Construct a rational expression:
% ================================
EX := (X+1)*(Y+1)*X*Y/Z;

       x*y*(x*y + x + y + 1)
ex := -----------------------
                z

FACTOR X;  EX;

  2
 x *y*(y + 1) + x*y*(y + 1)
----------------------------
            z

% Divide the denominator into each TERM:
ON DIV;  EX;

 2    -1                  -1
x *y*z  *(y + 1) + x*y*z  *(y + 1)

OFF DIV$
% Divide the denominator into each FACTOR
% (RAT ONLY works with FACTOR command):
ON RAT;  EX;

 2    y + 1          y + 1
x *y*------- + x*y*-------
        z              z

OFF RAT$
% Cannot simultaneously factor both x and y:
FACTOR Y;  EX;

  2 2    2         2
 x *y  + x *y + x*y  + x*y
---------------------------
            z

% (This is the same as just setting  OFF ALLFAC.)
% Finally tidy up by removing partial output factoring:
REMFAC X,Y$
```

If `OFF ALLFAC` had been set throughout then the common monomials in the *term groups* in the above expressions would not have been taken outside as factors.

An operator identifier can also be given as an argument to the `FACTOR` command, and then REDUCE will attempt to group all terms that contain that operator with identical arguments. (We give in §4.11 an example that "factors out" derivatives in an application of Leibniz' formula.)

Other relevant switches are `ON LIST`, which causes the terms in a sum to be output on separate lines, and `ON INTSTR`, which causes arguments of operators *and equations* to be output as they would if they were top-level expressions, i.e. taking account of the prevailing output format switch settings. (We give an example in §4.13 that allows us to control the format of differential equations and their solutions.) By default, arguments and equations are fully expanded, ungrouped and placed over a common denominator.

2.11 Operators

Operators are the most general objects available in REDUCE. They are usually parametrized, and can be parametrized in a completely general way. Only the operator identifier is declared in an `OPERATOR` declaration, thus:

`OPERATOR` *op1*, *op2*, ... ;

The number of parameters is *not declared*, and operators with the same name but different numbers of parameters are not algebraically related.

An operator needs to have at least one argument (or empty parentheses) to be recognized as an operator rather than a simple variable. [This is because REDUCE allows an identifier to be used as a simple variable and as an algebraically-unrelated operator (or some other types of parametrized objects) at the same time. To do so is generally not good practice because of its potential for confusion, but it does allow identifiers of such parametrized objects to be passed as parameters themselves, i.e. *passed* as simple variables and then *used* as parametrized objects. A similar facility is provided in most other languages, and an example of its use in REDUCE is provided in §9.2.2.]

REDUCE assumes that any unrecognized parametrized object is an operator; in batch mode it is automatically declared so, whilst in interactive mode the user is queried. Operators can be used to represent just about anything, although usually they will represent mathematical operators or functions.

By default, operators remain purely symbolic, as do variables, and can be considered as generalized or parametrized variables. Operators can also be given values, either for all values of all their arguments, or for restricted ranges, so that for example they can be used to represent discontinuous functions, or functions like the factorial function which might be considered to be defined only over the non-negative integers. Operators are usually given values by using let-rules, e.g.

```
OPERATOR COSEC, LN, QUADRATIC!-ROOT;

FOR ALL X LET COSEC(X) = 1/SIN(X),  LN(X) = LOG(X);
FOR ALL A, B, C, PM LET QUADRATIC!-ROOT(A, B, C, PM) =
      (-B + PM*SQRT(B**2 - 4*A*C))/(2*A);
```

although for fixed argument values assignments can also be used.

Operators are global in scope as are most objects in REDUCE, which means that once they have been defined they are available anywhere subsequently in the same REDUCE session. The declaration of an identifier to represent an operator is cleared by using CLEAR, but its actual values must all be cleared *first* (otherwise they persist, which seems to be a bug in CLEAR) thus:

```
FOR ALL X CLEAR COSEC(X), LN(X);  % do this FIRST then ...
CLEAR COSEC, LN;
```

It is possible to use SUB to substitute one operator identifier for another within an expression – see Rayna (1987), p. 53, for details. [However, note that Rayna's remarks in this context about arrays – see our §3.9 and §4.5 – do not apply to REDUCE 3.3.]

2.12 Omitting () and *

REDUCE makes a few concessions to the less explicit syntax that is usually used when writing mathematics on paper. For example, it is usual to write $\sin x$ rather than $\sin(x)$. REDUCE therefore allows *any* parametrized object that takes a single parameter to be written without the (), but the opening parenthesis "(" *must* be replaced by one or more spaces. Thus REDUCE also interprets SIN X as SIN(X), which it always outputs in the latter explicit form, although it would interpret SINX as a single identifier that has no connection with SIN. The rule is (informally) that REDUCE takes the first combination of objects that makes sense as the implied argument, which is the leftmost variable or prefix operator complete with its arguments, thus

```
SIN X;

sin(x)

SIN COS X;

sin(cos(x))

% But:
SIN X**2;

      2
sin(x)

% And beware!
SIN -X;

sin - x

% Should probably be
SIN(-X);

 - sin(x)
```

More precisely, by default prefix operators are applied before infix operators. However, if in doubt it is always best to use explicit `()`.

In view of the above convention it is not in general possible to use simple juxtaposition to imply multiplication, because it would not be clear whether `Y X` meant `Y*X` or `Y(X)`. However, an explicit number could not be an operator or other parametrized object, and so the * can be omitted from the form *number* * *identifier*[22]. The number can be in any explicit form, and need not be separated from the identifier, although it is safer if it is; for example, **2E** may cause trouble[23], presumably because it is confused with an incomplete real number specification, whereas

```
2 E;  % or 2 e

2*e
```

is clearly interpreted correctly.

One situation in which parentheses cannot simply be omitted is when invoking an operator without any parameters, because an empty parameter list is needed to force the identifier to be recognized as an operator invocation. A solution is to equate the operator to a variable with the same name, like this:

---

[22]New in REDUCE 3.3.

[23]With lower case **e** it crashes the current version of Atari REDUCE!

```
OPERATOR OP;  LET OP() = PI;
LET OP = OP();
OP;

pi
```

This example looks rather pointless, but the technique is very useful for allowing procedures, which behave much like operators, to be called like commands without any arguments. (However, algebraic procedure identifiers should not be declared to be operators.) For example, we have already seen that **WS** and **ED** can be used both with and without arguments.

2.13 Matrices

REDUCE understands matrix algebra, and algebraic operations can be performed on matrices exactly as they can on scalar objects, including multiplying and dividing matrices by scalars, and raising square matrices to positive and negative integer powers (the latter perform a matrix inversion). Matrices are the only parametrized objects in REDUCE that can be manipulated as composite objects without any parameters, as well as element-by-element. However, internally they are still manipulated element-by-element, so that before matrices can be used they must have had all their elements defined. Therefore, to manipulate what are mathematically matrices of indefinite size as purely symbolic objects, REDUCE matrices are not the appropriate representation, and non-commuting operators should be used instead (see §4.8.3).

The elements of a matrix M cannot be accessed until the dimensions or upper index bounds (the lower bounds are always 1) of M have been specified, which can be done either explicitly by a **MATRIX** declaration, or implicitly by assigning to M a matrix that is already defined; in the former case the elements all have the default value 0. Thus matrix elements cannot be purely symbolic any more than complete matrices can. However, purely symbolic objects can be assigned to matrix elements – they can be simply identifiers, or to preserve the flavour of matrix elements they can be unbound operator invocations, which can be made to look just like symbolic matrix elements, as illustrated below. The latter have another advantage, that they can be assigned in a loop statement – see §3.1.1.

REDUCE currently outputs matrices element-by-element, and not in the more customary two-dimensional tabular layout. This is because matrix elements can be arbitrary scalar expressions which might be very long and complicated, although a hybrid format could be used[24] as is done for rational expressions.

Here are some simple examples of matrix declaration, manipulation and output, edited to save space such that often only the first element of a matrix is shown and the rest are indicated by `etc`.

```
% Declaration:
MATRIX M(2,2);

% Assignment and display:
% Note that declared matrix M is zero by default.
MM := M;

mm(1,1) := 0
mm(1,2) := 0
mm(2,1) := 0
mm(2,2) := 0


% Assignment of an operator to the elements:
% Note that here in batch mode REDUCE assumes el to be
% an operator, but would query this in interactive mode.
M(1,1) := EL(1,1)$  M(1,2) := EL(1,2)$
M(2,1) := EL(2,1)$  M(2,2) := EL(2,2)$

*** el declared operator

% Anonymous display:
M;

mat(1,1) := el(1,1)
mat(1,2) := el(1,2)
mat(2,1) := el(2,1)
mat(2,2) := el(2,2)


% Named display using a "null assignment":
M := M;

m(1,1) := el(1,1)
etc.


% Matrix algebra - note that a and b are scalars:
MM := A*M**2;
                                            2
mm(1,1) := a*(el(2,1)*el(1,2) + el(1,1) )
etc.
```

---

[24] We understand that improved matrix output will be provided in future versions of REDUCE, but in the meantime see §9.2 for a very crude tabular matrix printing routine.

```
MM := M/B + MM;
                                                 2
              el(2,1)*el(1,2)*a*b + el(1,1) *a*b + el(1,1)
mm(1,1) := ---------------------------------------------
                                   b
etc.

MM := M * MM;

mm(1,1) := (el(2,2)*el(2,1)*el(1,2)*a*b + 2*el(2,1)*

                                                     3
   el(1,2)*el(1,1)*a*b + el(2,1)*el(1,2) + el(1,1) *a*b +

           2
   el(1,1) )/b
etc.

MM := M**(-1);

                          el(2,2)
mm(1,1) := -----------------------------------
              el(2,2)*el(1,1) - el(2,1)*el(1,2)
etc.
```

The "anonymous display", in which a matrix is called `mat`, is unavoidably used to display matrix-valued expressions that have no name. However, when the value of a specified matrix is being displayed, as illustrated in the middle of the above example, REDUCE ought to use its name automatically. This can be forced by using a "null assignment", also illustrated above, but we hope this will become unnecessary in future versions.

The operator **LENGTH** applied to a matrix returns its dimensions or upper index bounds as a list. REDUCE matrices are *always* two-parameter objects, so that row or column matrices must be declared to have respectively a row or column dimension of 1, and elements must always be accessed using two indices[25], even if one of them must be 1. Matrix elements can appear anywhere that scalar variables can appear. There is a more succinct way of assigning to *all* the elements of a matrix than explicitly assigning to each element as above, and this uses the operator **MAT**. The arguments to **MAT** are "rowlists", which are sequences enclosed in ordinary round parentheses of comma-separated values to be assigned to the elements in a row. One can use REDUCE's

[25]It is not too hard to remove this restriction, and we have written a small extension to the standard matrix code to provide some "syntactic sugar" to support row and column matrices more naturally.

input flexibility to lay out the arguments to MAT so as to make it look like a matrix, e.g.

```
MATRIX M;

*** matrix m redefined

M := MAT ( (A, B) ,
           (C, D) );

m(1,1) := a
m(1,2) := b
m(2,1) := c
m(2,2) := d
```

Note that the declaration MATRIX M is optional, i.e. redundant. The operator MAT defines an anonymous matrix that can just as well be used within an expression as assigned to a matrix variable, and it is essentially what is used internally to represent the value of a matrix.

There are no predefined special matrices. A null matrix is created just by declaring a matrix with specified dimensions, and the identity or unit matrix can be established by assigning 1 to the diagonal elements, using a loop construct (see §3.1.1) if the matrix is large[26]. The operator TP returns the transpose of a matrix, i.e. the matrix with rows and columns swapped; for example, it turns a row matrix into a column matrix:

```
TP MAT((1,2,3));

mat(1,1) := 1
mat(2,1) := 2
mat(3,1) := 3
```

Note that this example illustrates a way of inputting column matrices with a minimum of parentheses, although it is possible to provide better support for row and column matrices – see §G.

Matrices of course need not be square, but if they are then there are some special operations that can be applied to them. One is inversion, which is effected by raising to the power (-1) as we have seen above. This provides a way of solving coupled linear equations, but unless the system is square and non-singular it is probably more convenient to use the operator SOLVE. In addition, the operators DET and TRACE return respectively the determinant and the trace (i.e. the sum of the diagonal elements), thus:

---

[26]See also §9.2. Internally there is a routine to generate the identity matrix of a given size that can be fairly easily made available in algebraic mode – see the Appendix, §G.

```
M := MAT ( (A, B) ,
           (C, D) )$
DET M;

a*d - b*c

TRACE M;

a + d
```

A very common matrix problem is to find the eigenvalues and eigenvectors of a square matrix. The operator **MATEIGEN(*matrix*, *id*)**[27] attempts to solve this problem by first constructing the characteristic (eigenvalue) polynomial using the identifier *id* (which must be unbound) as the unknown, i.e. the eigenvalue. It then attempts to factorize this characteristic polynomial into square-free[28] factors and returns the result as a list of three-element lists of the form

{ {*square-free factor*, *multiplicity*, *eigenvector*} ... }

where the eigenvector corresponding to each square-free factor of the characteristic equation is represented as a column matrix. A simple example should make this clearer:

```
% Establish general 2*2 matrix (no need for declaration):
M := MAT((A,B), (C,D));
% Construct characteristic equation and eigenvector(s):
EV := MATEIGEN(M, LAM);

                                              2
ev := {{a*d - a*lam - b*c - d*lam + lam ,

         1,

                           arbcomplex(1)*b
         mat(1,1) :=  -  -----------------
                               a - lam

         mat(2,1) := arbcomplex(1)}}

CEQN := FIRST FIRST EV;

                                              2
ceqn := a*d - a*lam - b*c - d*lam + lam

EVEC := THIRD FIRST EV;
```

[27]New in REDUCE 3.3, still experimental and not yet completely reliable.

[28]i.e. non-multiple – defined properly in chapter 7.

```
                          arbcomplex(1)*b
evec(1,1) :=  - -----------------
                             a - lam

evec(2,1) := arbcomplex(1)

% Establish the simplification rule that ceqn = 0,
% cheating slightly because let-rule arguments are
% not evaluated when the rule is defined:
LET A*D = A*D - CEQN;

% Check explicitly that m*evec = lam*evec (symbolically):
M*EVEC;

                       arbcomplex(1)*b*lam
mat(1,1) :=  - ---------------------
                             a - lam

mat(2,1) := arbcomplex(1)*lam

LAM*EVEC;

                       arbcomplex(1)*b*lam
mat(1,1) :=  - ---------------------
                             a - lam

mat(2,1) := arbcomplex(1)*lam

% and just to be 100% sure:
ZERO := M*EVEC - LAM*EVEC;

zero(1,1) := 0
zero(2,1) := 0
```

Note that the let-rule

```
LET <<CEQN>> = 0;
```

is a more elegant version of the let-rule used above, but uses a facility that is not currently advertised, and cannot be used when clearing let-rules – see §3.7.1.

Matrix variables, like operators, are global in scope (see §3.7.4), although the internal representation of a matrix is a local object, so that with care and a little "inside" information matrices can be manipulated as local objects – see §9.2. It is to be hoped that future versions of REDUCE will provide proper support for local matrices.

Let-rules work to some extent for matrix *identifiers*, e.g. if `M` is a matrix identifier then `LET MM = M` performs a partial assignment of the structure of `M` to `MM`, in that `MM` then evaluates to `M` but cannot be directly indexed, i.e. "`MM;`" works but "`MM(1,1);`" fails. We do not recommend this usage. (See also Rayna (1987), p. 56.)

2.14 Exercises

1. If the vectors $\vec{a}$ and $\vec{b}$ are expressed in terms of the Cartesian basis vectors $\vec{\imath}, \vec{\jmath}, \vec{k}$ as $\vec{a} = 2\vec{\imath} - \vec{\jmath}$ and $\vec{b} = \vec{k}$, calculate the angle between $\vec{a}+\vec{b}$ and $\vec{a}-\vec{b}$. [Hint: Represent the vectors by simple REDUCE variables, regard algebraic multiplication (*) applied to the "basis vectors" `i`, `j`, `k` as the scalar product operation, assign values to their 6 distinct scalar products, and proceed interactively in the simplest possible way. It's surprisingly easy!]

2. Write REDUCE `SUB` statements whose values are

   (a) the complex conjugate,

   (b) the real part,

   (c) the imaginary part,

   of a complex-valued quantity z (of the form $z = u + iv$, where u and v are real). Test them on general expressions including $1/(x+iy)$. [Hint: Use the generally valid expressions for the real and imaginary parts of a quantity in terms of the quantity and its complex conjugate.] Use this and de Moivre's theorem to find the expansion of $\cos 7x$ in terms of powers of $\cos x$ and $\sin x$.

3. Establish let-rules that will recursively replace $\cos nx$ and $\sin nx$ by the sums of products of $\sin x$, $\cos x$, $\sin(n-1)x$ and $\cos(n-1)x$ corresponding to the decomposition $nx = x + (n-1)x$ for all $n > 1$. Hence check the expansion of $\cos 7x$ obtained above.

4. If $x = 3\cos\theta - \cos 3\theta$ and $y = 3\sin\theta - \sin 3\theta$ find d^2y/dx^2 in terms of functions of θ in a compact form (i.e. as completely factorized as possible and containing the smallest possible number of explicit terms).

5. Obtain the equation of the tangent to the curve $y = x^3$ at $x = 1$, and hence find the coordinates of the point where the tangent cuts the curve again.

6. Show that the graphs of $y = 3e^x$ and $y = \cosh x$ intersect at $x = -\frac{1}{2}\ln 5$, and calculate the area of the finite region bounded by these two curves and the y axis.

   Find the volumes generated when this area is rotated through 360° about (a) the x axis and (b) the y axis. [Hint: The latter is best expressed as an integral over infinitesimal *cylinders*, which gives a simpler integral in x than the more usual integral in y arising from integrating over infinitesimal discs.]

[Remember that let-rules for `EXP(X)` and `SQRT(X)` need to be established in terms of `E**X` and `X**(1/2)`. Also, the integrator *appears* to clear let-rules, which work again if re-set. In fact, what happens is that the integrator returns its results with a marker that shows they have already been simplified (see §6.2), so that the results are not re-simplified; the reason is that otherwise REDUCE would repeatedly attempt to perform any integration that failed and so was returned in symbolic form. Re-setting a let-rule causes all such markers to be cleared, but so does directly calling the internal procedure `SYMBOLIC RMSUBS()`, as discussed in §6.2. An alternative way to reset these markers, using purely algebraic mode, is to change a suitable but harmless switch setting, e.g. "`ON FLOAT; OFF FLOAT;`" will work. In REDUCE 3.3 an expression like `5/SQRT(5)` is simplified to `SQRT(5)` automatically if the switch `RATIONALIZE` is on, but in previous versions there appears to be no way to *automatically* effect this simplification – not that it is worth worrying about!]

7. Find by explicit computation the number of distinct integer powers of the following matrices:

$$\begin{pmatrix} 0 & 1 \\ 1 & 0 \end{pmatrix}, \quad \begin{pmatrix} 0 & 1 \\ -1 & 0 \end{pmatrix}, \quad \begin{pmatrix} 0 & i \\ -1 & 0 \end{pmatrix}.$$

[This example relates to group representations.]

8. Use matrix inversion to solve for x, y the pair of simultaneous equations with symbolic coefficients

$$\begin{aligned} ax + by &= c, \\ dx + ey &= f. \end{aligned}$$

9. If

$$j_x \equiv \begin{pmatrix} 0 & 1 \\ 1 & 0 \end{pmatrix}, \quad j_y \equiv \begin{pmatrix} 0 & -i \\ i & 0 \end{pmatrix},$$
$$j_z \equiv \begin{pmatrix} 1 & 0 \\ 0 & -1 \end{pmatrix}, \quad u \equiv \begin{pmatrix} 1 & 0 \\ 0 & 1 \end{pmatrix},$$

confirm by explicit computation that $j_\ell^2 = u$ for $\ell = x, y, z$ and

$$j_\ell j_m - j_m j_\ell = 2i j_n, \quad \ell, m, n \text{ cyclic.}$$

[This example relates to Pauli spin matrices in quantum mechanics, or equivalently to the Lie group $SU(2)$.]

10. Using **MATEIGEN** and **SOLVE**, or otherwise, find the eigenvalues and eigenvectors of the two-dimensional rotation matrix

$$R(\theta) \equiv \begin{pmatrix} \cos\theta & \sin\theta \\ -\sin\theta & \cos\theta \end{pmatrix}.$$

[Note that they are complex.]

11. Show that the function $z = (x^2+xy+y^2)\exp(-x^2)$ has stationary values at the points with coordinates $(0,0)$, $(-1,\frac{1}{2})$, $(1,-\frac{1}{2})$, and by finding the determinant of the Hessian matrix (cf. §4.5), or otherwise, determine whether they are maxima, minima or saddle points.

12. Calculate the Jacobian factor of the coordinate transformation from (x,y) to (u,v) defined by $u = x^2/y$, $v = y^2/x$. [Don't try too hard to make REDUCE do all the work!] Hence evaluate $\int_R x^2y^2\,dxdy$ taken over the region R bounded by the curves $y^2 = x$, $y^2 = 2x$, $x^2 = y$, $x^2 = 2y$.

3

REDUCE as an algebraic programming language

A REDUCE "program" is not a well-defined object like a Pascal or FORTRAN program, but is just a body of code intended to perform a particular task. It will usually be a collection of procedures (equivalent to procedures, functions and subroutines in other languages), control structures, let-rules and variable assignments. Often these will be read or loaded from a file, and then the "program" will be run by invoking a procedure, which may itself call other procedures. Alternatively, the "program" may run as it is being read just as it would if the statements were entered one-by-one from the keyboard.

Thus the distinction between REDUCE as an interactive algebraic calculator and as a programming language is an artificial one (as is the distinction between calculator and computer), and most REDUCE sessions will contain elements of both. Nevertheless, a good approach to computing is that of tool building: whenever a problem has to be solved, a body of code is developed that will solve a general version of the problem, and it is then added to a collection of such general-purpose tools. In REDUCE such tools are usually best packaged as procedures, which will be introduced in the next chapter. This chapter discusses the necessary ingredients such as conditional statements, loop statements and composite statements, and introduces an additional structured data type, the array.

3.1 Control structures

An essential feature of "programming" is the use of control structures, that is, mechanisms for automatically controlling the execution of the program. REDUCE provides those found in other modern languages, and although REDUCE was actually modelled on ALGOL 60 its control structures are very similar to those in Pascal (which was developed from ALGOL 60). The control structures all control code that is syntactically a *single statement*, but as will be explained in §3.7 a block of statements can be wrapped up so as to constitute syntactically a single statement, very much as in Pascal.

Careful indentation of subordinate code in control structures is a very good way of making the logical structure of a program clear, as (we hope) is illustrated by our subsequent longer examples. We have followed the rules used by Anthony Hearn in the REDUCE source, which are that each new level of the code is indented by an additional 3 spaces (and that any descriptive comment follows rather than precedes the first line of a procedure definition – see the next chapter).

3.1.1 Iteration over a predetermined numerical range

One of the simplest programming tasks is to perform some operation repeatedly in what is called a *loop*, and the simplest version of this is to execute the code in the loop a *predetermined* number of times. This implies using a variable to hold a count of the number of times the loop has been executed, and the value of this variable is made available within the body of the loop; it need not be used although usually it is. In REDUCE the syntax of this simple kind of loop is

FOR *id* := *start* STEP *inc* UNTIL *stop* DO *statement*

where the keywords are shown in capitals. The variable *id* is automatically declared to be an integer local to the loop (unlike in most other languages); its value can be used within the statement controlled by the loop, but it should not be changed. *Start*, *inc* and *stop* should be either integer constants or variables with integer values; they are allowed to be general expressions, but they are re-evaluated each time round the loop causing either gross inefficiency or potentially undesired side-effects if they involve function calls. This is a feature of ALGOL 60 that has subsequently been abandoned as bad practice, and the REDUCE semantics ought to be updated to reflect this[1]. (However, in the meantime it does provide the only way of exiting the loop early without making an explicit jump.) These idiosyncrasies are illustrated at the end of §3.7.1.

For-loops are often used to access parametrized objects, as in this simple example:

```
% Construct a unit matrix:
matrix m(5,5);
for i := 1 step 1 until 5 do m(i,i) := 1;
```

[1]We understand that, in fact, they will.

Because the construct "STEP 1 UNTIL" is used so often it can be abbreviated to a colon, as shown in the following example of a double loop:

```
% Construct a matrix with symbolic entries:
matrix m(5,5);  operator mm;
for i := 1 : 5 do  for j := 1 : 5 do  m(i,j) := mm(i,j);
```

The above form of FOR statement has no algebraic value[2] and so produces no output. However, REDUCE allows the keyword DO to be replaced by either SUM or PRODUCT, thereby producing a succinct structure for generating a repeated sum (Σ) or repeated product (Π) respectively, which has no analogue in conventional languages. In this case it is essential that the statement controlled by the loop construct has a value, and in many cases it will just be an expression, e.g.

```
% A truncated power-series approximation:
off allfac;  on div, revpri;
LOG1PLUSX := FOR N := 1 : 5 SUM X**N/N;

                   1  2    1  3    1  4    1  5
log1plusx := x + ---*x  + ---*x  + ---*x  + ---*x
                  2        3        4        5

on allfac;  off div, revpri;

% A large number:
FACTORIAL100 := FOR I := 1 : 100 PRODUCT I;

factorial100 :=
9332621544394415268169923885626670049071596826438162146859
2963895217599993229915608941463976156518286253697920827223
758251185210916864000000000000000000000000
```

3.1.2 Iteration involving lists

There are also two other replacements for the keyword DO that build lists as the values of FOR statements: the keyword COLLECT collects the values of the controlled statement into a list; the keyword JOIN assumes that the value of the controlled statement is a list, and appends these lists together, as in the following simple examples:

```
FOR I := 1 : 5 COLLECT I**2;

{1,4,9,16,25}

FOR I := 1 : 5 JOIN {I**2};

{1,4,9,16,25}
```

[2]It has the Lisp value NIL.

This provides a mechanism for converting a matrix into a list:

```
% Construct an arbitrary matrix:
NROWS := 4$  NCOLS := 3$
MATRIX M(NROWS, NCOLS);  OPERATOR EL;
FOR I := 1 : NROWS DO  FOR J := 1 : NCOLS DO
      M(I,J):=EL(I,J);

% Convert it into a list of lists:
ML := FOR I := 1 : NROWS COLLECT
      FOR J := 1 : NCOLS COLLECT M(I,J);

ml := {{el(1,1),el(1,2),el(1,3)},
       {el(2,1),el(2,2),el(2,3)},
       {el(3,1),el(3,2),el(3,3)},
       {el(4,1),el(4,2),el(4,3)}}
```

When the purpose of a loop is to perform some operation on each element of a list there is an elegant construct available that provides all the power of the for-loop, and has the general form

FOR EACH *id* IN *list* DO *statement*

where FOR EACH can also be written as FOREACH. (This construct is also particularly useful in its symbolic mode variant – see the end of §6.3.) As with the for-loop, the keyword DO can be replaced by SUM, PRODUCT, COLLECT or JOIN. The identifier is automatically declared to be a scalar variable local to the body of the loop, and *list* is an expression which is assumed to evaluate to a list. Here are some simple examples:

```
% Reverse a list (without using reverse!):
THELIST := {A, B, C, D}$
REVLIST := {}$
FOR EACH EL IN THELIST DO REVLIST := EL . REVLIST;
REVLIST;

{d,c,b,a}

FOR EACH X IN {A, B, C} SUM X**2;

 2    2    2
a  + b  + c

FOR EACH X IN {A, B, C} JOIN {X**2};

  2  2  2
{a ,b ,c }

FOR EACH X IN {A, B, C} COLLECT {X**2};

   2    2    2
{{a },{b },{c }}
```

We now have two ways to convert a list into a matrix:

```
THELIST := {A, B, C}$

MATRIX M(1,3);

% Either:
J := 0$
FOR EACH EL IN THELIST DO M(1, J:=J+1) := EL;
M;

mat(1,1) := a
mat(1,2) := b
mat(1,3) := c

% or:
THELIST := 0 . THELIST;

thelist := {0,a,b,c}

FOR I := 1 : 3 DO
   M(1,I) := FIRST (THELIST := REST THELIST);
M;

mat(1,1) := a
mat(1,2) := b
mat(1,3) := c
```

The first is slightly more elegant. Whenever iterating through the elements of two or more different kinds of structured objects together (here a matrix and a list) it is often sufficient to use one of the structures explicitly to control the loop; which to use is a matter of taste and the precise context. In the above two examples we have used the matrix to control the iteration in the first and the list to control the iteration in the second.

3.2 Conditions and logical expressions

More general loop constructs and other kinds of conditional execution require the use of conditions that are either true or false, i.e. Boolean, logical or logic-valued expressions. In algebraic mode, conditions or logical expressions can only be used *immediately* in tests; there are no logical constants or variables, and the logical value of all non-logical expressions – even 0 – is true. This is not a major constraint because even in conventional languages that do have logical constants and variables they are not used very often, and it is always possible to simulate their use.

3.2.1 Relational infix operators

The result of comparing two quantities or objects is a logical value, and the relational infix operators that perform such comparisons have the same symbols as in most languages (with the idiosyncratic exception of versions of FORTRAN before FORTRAN 90). They are

```
= NEQ > >= < <=
```

where the only one that varies much between languages is NEQ, which represents "not equal" ($\neq$). Arbitrary scalars can be compared for equality or inequality, but only numerical values (real, rational or integer – not complex) can be compared for relative magnitude. Therefore, if any of the expressions being compared involve functions other than rational polynomials it will be necessary to set ON NUMVAL together with a floating-point number domain (see the example below in §3.4).

3.2.2 Predefined predicates or logic-valued functions

These are mostly prefix operators, i.e. they are used like ordinary functions:

- NUMBERP X is true if X evaluates to any type of explicit number (except complex unless COMPLEX is ON);
- FIXP X is true if X evaluates to an integer;
- EVENP X is true if X evaluates to an even integer[3];
- FREEOF(Y,X) or Y FREEOF X is true if the expression Y does not contain the kernel X explicitly anywhere in its structure and has not been declared to DEPEND directly[4] on X.

3.2.3 Logical operators

The logical infix operators AND and OR, and the logical prefix operator NOT, operate on logical values such as those returned by the above relational operators or predicates. For example, the mathematical condition $0 < x \leq 1$ must always (as is generally the case in programming languages) be written in the more explicit form

---

[3]New in REDUCE 3.3.

[4]There are various depths of indirect dependence that it might be appropriate to test, of which the supplied version of FREEOF takes a particular view – a much more general version that is still in the same spirit is presented in the Appendix, §D.

```
0 < X  AND  X <= 1
```

Evaluation of expressions involving **AND** or **OR** is strictly left-to-right, and stops as soon as the overall value of the expression is determined. Therefore they can be used to express certain forms of conditional execution succinctly: for example, a test of relative magnitudes of quantities, which would give rise to an error if the quantities were not numerical, can be preceded by tests that ensure that they are numerical. Thus a very common kind of logical expression has the form

```
NUMBERP X  AND  NUMBERP Y  AND  X > Y
```

The rules for combining relational and logical infix operators follow the natural precedence order, so that the above examples are interpreted correctly without parentheses (unlike in Pascal). **NOT** is a prefix operator, and so has higher precedence than *any* infix operator, whilst **AND** has higher precedence than **OR**. Parentheses `( )` can be used as usual within logical expressions to make the order of evaluation explicit, and it is probably better in general to use parentheses than to rely on the rules of precedence.

In the following, a logical expression constructed as described above will be referred to as a *condition*.

3.3 General loops: WHILE and REPEAT

The general loop is controlled by a more general condition than the value of an iteration counter, and REDUCE has two such constructs:

`WHILE` *condition* `DO` *statement*
`REPEAT` *statement* `UNTIL` *condition*

These two constructs are the same as in Pascal, except that for both of them the controlled statement must be syntactically a single statement, whereas Pascal allows a statement-sequence in a repeat-loop.

There are two differences between while- and repeat-loops. Firstly, in a while-loop the condition is tested before the loop body is ever executed, whereas in a repeat-loop the loop body is always executed at least once. (This is similar to the difference between the do-loop in FORTRAN 77 and FORTRAN 66 respectively.) It is therefore essential to ensure that the condition is properly initialized before executing a while-loop, whereas in a repeat-loop the condition can be computed within the loop body without prior initialization. Secondly, the condition in a while-loop is a continuation condition, whereas in a repeat-loop it is a termination condition, one being the negation of the other. Neither of these loop constructs ever returns an algebraic value. Examples of them will be given later (WHILE in §§3.8 and 4.11 and REPEAT in §4.14).

3.4 Conditional statements: IF

Simple conditional execution of a statement is achieved using

IF *condition* THEN *statement*

and alternative execution of two statements is achieved using

IF *condition* THEN *statement1* ELSE *statement2*

The statement following THEN is executed if the condition is true, otherwise the statement following ELSE is executed if there is one. In either case execution then proceeds with the statement following the whole IF construct. The value of the IF statement is the value of the controlled statement that is executed, and if no statement is executed then the IF statement has no algebraic value. Note that there must be no terminator (";" or "$") before an ELSE because the terminator would terminate the whole IF construct; including one is a common mistake.

The use of the *value* of the IF statement is unfamiliar from conventional languages, but can be quite useful; an example that makes serious use of this feature is the definition of the operator SGN in §3.6. The following two code fragments produce the same result, although the output that they produce is slightly different:

```
THETA := 3/2$
ON NUMVAL, FLOAT;  % or bigfloat
IF SIN THETA > THETA THEN X := 1 ELSE X := -1;

-1

X := IF SIN THETA > THETA THEN 1 ELSE -1;

x := -1
```

The statements controlled by a conditional statement must be syntactically single statements, but that allows them to be conditional statements so that nested conditional statements can be constructed. There is no "elseif" kind of keyword as there is in some languages (such as FORTRAN 77) because there is no need. There is also no "case" statement in REDUCE (for which there probably is a need), and the best simulation is a nested conditional statement of the form:

```
     IF x = val1a OR x = val1b ... THEN statement1
ELSE IF x = val2a OR x = val2b ... THEN statement2
ELSE IF ...
ELSE default-statement;
```

Note that there must not be any terminators within this construct (apart from any buried inside the controlled statements). This construct is actually much more general than the Pascal case statement, in that it has a default clause to catch the case that the control variable `x` does not have any of the "expected" values, and the conditions tested could be more general than those shown above. Also this construct can have a value, which is the value of the controlled statement that is executed. The construct shown above is apparently ambiguous; the rule for interpreting it is that an `ELSE` associates with the *nearest* `IF` (that has no `ELSE`) unless parentheses `( )` or one of the block structures described later is used to make the required association explicit. A mistake that is very easy to make and very hard to spot is incorrect association of `ELSE` with `IF`. This is one situation in which indenting can make code appear correct when it is not, and explicit grouping is always safest. Parentheses should be used for this on grounds of efficiency unless a block structure is necessary for other reasons.

Although the values of logical expressions cannot be output directly in algebraic mode as can those of other types of expression, they can still be output indirectly as in the following example:

```
% if condition then true else false; e.g.
IF 2 > 1 THEN TRUE ELSE FALSE;

true
```

Here `TRUE` and `FALSE` are two variables whose identifiers have no meaning to REDUCE itself, only to us. We used this trick earlier to show that with the switch setting `OFF EXP` equal expressions may not compare equal, without explaining it at the time.

3.5 Writing new predicates

This is possible, but only in symbolic mode (which will be discussed in chapter 5 et seq.) as illustrated in the following simple example, which we include at this point for completeness:

```
symbolic procedure oddp n; fixp n and not evenp n;
symbolic operator oddp;
% and as a test from algebraic mode:
IF ODDP 33 THEN ODD ELSE EVEN;
```

```
odd

IF ODDP 32 THEN ODD ELSE EVEN;

even
```

Algebraic operators cannot be logic valued, so the above procedure definition cannot be replaced by a let-rule.

3.6 Conditional let-rules and partially evaluated operators

As mentioned earlier, the most general form of let-rule is

FOR ALL *varlist* SUCH THAT *condition* LET *rule*

The use of conditional let-rules nearly always involves operators, and allows the values of operators to be defined in very general ways. For example, many operations can only be performed on numerical values, but any expression may become numerical later in a calculation. So operators that require numerical arguments should be defined so that they remain symbolic and "wait" until their arguments become numerical. A good example is a sign function, which could be defined as follows:

```
OPERATOR SGN;
FOR ALL X SUCH THAT NUMBERP X LET SGN X =
   IF X > 0 THEN 1 ELSE IF X < 0 THEN -1 ELSE 0;
SGNA := SGN A;

sgna := sgn(a)

SGN 123;

1

SGN (-3/4);

-1

% Finally make A numerical:
A := 0$  SGNA := SGNA;

sgna := 0
```

REDUCE allows recursion (as do most modern languages except versions of FORTRAN before FORTRAN 90), that is, an operator or procedure can be defined in terms of itself. When called, the operator or procedure will call itself repeatedly, and some condition is needed to stop the process. A conditional let-rule provides a way of defining recursive operators; here is an example that reduces the argument of the SIN operator to a standard range of $(-\pi, \pi]$ if it has the form of an explicit multiple of π:

```
FOR ALL N SUCH THAT NUMBERP N AND N > 1 LET
   SIN (N*PI) = SIN ((N-2)*PI);
% REDUCE already knows about multiples of
% pi, pi/2, pi/3 and pi/4, so ...
% The above simple let-rule needs
ON RATIONAL$
SIN (17 PI/5);

        3
 - sin(---*pi)
        5

SIN (11 PI/7);

        3
 - sin(---*pi)
        7
```

REDUCE allows values to be defined for built-in operators in the same way as user-defined operators, so that new differentiation rules can be defined. For example, the derivative of the built-in absolute value function could be defined as follows. This example requires a little care, because differentiation with respect to a number is not defined, so the most general approach is to go via a subsidiary operator, thus:

```
Y := DF(ABS X, X);

y := df(abs(x),x)

SUB(X = 1, Y);

sub(x=1,df(abs(x),x))

% This has not been evaluated usefully, so define ...

OPERATOR STEP;
FOR ALL X LET DF(ABS X, X) = STEP X;
FOR ALL X SUCH THAT NUMBERP X LET STEP X =
   IF X > 0 THEN 1 ELSE IF X < 0 THEN -1 ELSE UNDEFINED;

Y;

step(x)

SUB(X = 1,  Y);

1
```

```
SUB(X = -5, Y);

-1

SUB(X = 0,  Y);

undefined
```

3.7 Composite statements: groups, blocks and local variables

There are two kinds of composite statement in REDUCE called a *group* and a *block*, which provide ways of grouping statements into units that are syntactically equivalent to a single statement. They are therefore particularly useful in control structures and as procedure bodies, but they can also be useful elsewhere, such as on the right of let-rules or assignments and occasionally where local variables are needed in the middle of a piece of code. The values of statements used within composites are *never output* (although they may be used), so it makes no difference whether they are terminated by ";" or "$". Which to use is a matter of taste, but for consistency with other languages we use ";".

It is worth remarking here that what we have called *terminators* are really mainly used as *separators*, and the rules governing exactly where they should and should not appear can be confusing, as they can in Pascal and most other ALGOL-based languages – only C consistently uses ";" as a *terminator*. In situations where statements are enclosed in parentheses, e.g. to force the correct association of clauses in `IF ... ELSE` constructs, or when they are separated by commas, e.g. when used as arguments, they *must not* be terminated.

Blocks in REDUCE share many of the features of blocks in other languages, but have some potentially confusing differences, because they are really Lisp "prog" constructs (see §6.6). One consequence of this is that the REDUCE `RETURN` statement is a Lisp `RETURN` and is unlike the return statement in almost any other language – it "returns" from a block, and has nothing directly to do with returning from a procedure; to put it another way it *returns a value* rather than returning to a particular location in the program.

3.7.1 The group statement

In many cases it is simply necessary to group a number of statements together and possibly to return a value, which is often naturally the value of the last statement in the group. For this, REDUCE provides the *group statement* delimited by the symbols "<<" and ">>" thus:

<< *stmt1* ; *stmt2* ; ... ; *stmtn* >>

This construct simply executes the statements *stmt1*, *stmt2*, ... in sequence, and the value of the group statement is the value of the last statement, *stmtn, which must not be terminated.* If the last statement is terminated, then there is a non-terminated null statement after it, and it is the value of that null statement that is returned as the value of the group. Hence the last statement should be terminated only to *force* the group to have no algebraic value, which is unlikely to be necessary since the value of any statement (including a composite) need not be used. A group statement is not suitable if it is necessary to return the value of an intermediate statement as the value of the composite, or if more complicated code needs to be executed within the composite.

As an example of the use of the group statement, let us return to our earlier example of two ways to convert a list into a matrix, and do them differently, the second one more elegantly:

```
THELIST := {A, B, C}$
MATRIX M(1,3);
% Either:
J := 1$
FOR EACH EL IN THELIST DO <<M(1,J) := EL; J:=J+1>>;
% or:
FOR J := 1 : 3 DO <<
   M(1,J) := FIRST THELIST;
   THELIST := REST THELIST >>;
```

These examples did not use the value of the group, but the following modification of the first example does (trivially):

```
J := 1$
LISTLENGTH := FOR EACH EL IN THELIST SUM
   << M(1,J) := EL; J:=J+1; 1 >>;

listlength := 3
```

A side-effect of the fact that a group statement evaluates to the value of its last statement is that it can be used to force evaluation in situations where this is not otherwise possible, as illustrated in the following three examples:

```
EQN := (X = A);  SOLVE <<EQN>>;

FILE := "TEST.RED";  IN <<FILE>>;

D := 3;  LET X**<<D>> = 0;
```

It could be argued that `SOLVE` and `IN` ought to evaluate their arguments, and perhaps in later versions of REDUCE they will, but there are good reasons for `LET` not to do so by default, and this use of << >>

to force evaluation of parts or all of the left side of a let-rule can be very useful, as we have already illustrated at the end of chapter 2.

Another very useful application of `<< >>` in the same spirit is as "superparentheses" (Rayna 1987, p. 112), to force components of expressions to be fully evaluated in a way that normal parentheses `( )` do not, as illustrated by the following example:

```
FOR ALL X LET X**2 = 1;  LET A*B = 0;

A*B/( (A*A) * (B*B) );

***** zero denominator


A*B/(<<A*A>>*<<B*B>>);

0
```

In REDUCE 3.3 it is also possible, but undocumented, to use "`{ }`" instead of "`<< >>`"; the former construct is interpreted as an algebraic-mode list if the statements in it are separated by commas, and as a group if they are separated by semicolons or dollars.

Finally, here is a demonstration of the idiosyncratic way in which the control variables of the `FOR`-loop can be changed dynamically in REDUCE[5]. This is not a generally recommended programming style, but is included to illustrate what can be done – possibly by mistake! Because this is not a simple iteration it should be implemented as one of the more general loop constructs rather than a corrupted `FOR`-loop.

```
J := 1$  K := 1000000000$
FOR I := 1 STEP J UNTIL K COLLECT
   << IF (J := I) > 100 THEN K := 0; I >>;

{1,2,4,8,16,32,64,128}
```

3.7.2 The block or compound statement

The more general composite statement is called a *block* or *compound* statement and is delimited by the statements `BEGIN` and `END` familiar from other ALGOL-based languages. The syntax of the block statement looks superficially very similar to the group statement, and is

`BEGIN` *stmt1* ; *stmt2* ; ... ; *stmtn* `END`

A terminator before the `END` is optional but unnecessary, and regardless of this terminator a block statement has by default no algebraic value.

---

[5]At least in versions of REDUCE up to 3.3.

To give a value to a block statement it is necessary to use a RETURN statement within the block, which consists of the keyword RETURN followed immediately (with no terminator) by any statement, the value of which becomes the value of the block. A RETURN statement in which the keyword RETURN is followed immediately by a terminator or END returns no algebraic value. As an example, the following two composite statements are entirely equivalent, although the group statement is more succinct and may execute slightly faster, and so is preferable:

```
<< stmt1 ; stmt2 ; ... ; stmtn >>
BEGIN stmt1 ; stmt2 ; ... ; RETURN stmtn END
```

However, RETURN may appear anywhere in a block statement, and has the additional effect of terminating execution of the block, so that if the RETURN is controlled by a conditional statement then alternative returns are possible. Simple uses of blocks are illustrated quite frequently in the rest of this book; the first explicit example appears in §3.7.5.

Composite statements can be nested, and a RETURN relates to the smallest enclosing block statement. The RETURN may be inside a group inside a block, and causes the whole block statement to return. However, a RETURN cannot appear within a loop construct (FOR, WHILE or REPEAT), and therefore some programming models frequently used in other languages are not available in REDUCE. In these situations (but in few others) it is necessary to program at a more explicit level using the GOTO statement and labels.

3.7.3 The GOTO statement and labels

These can be used *only* within block (or compound) statements. A label is any identifier (*not* a number) followed by a colon, and can appear only at the top level within a block statement. Labels are not explicitly declared. The statement GOTO also has the synonym GO TO, and it can appear alone or within a conditional statement. Hence, for example, the while-loop

```
WHILE condition DO statement
```

can be simulated by

```
BEGIN
   start:
   IF condition THEN << statement; GOTO start >>
END
```

or, slightly less elegantly, by

```
BEGIN
   start:
   IF NOT condition THEN RETURN; statement; GOTO start
END
```

Both these compound versions have the advantage that the body statement can contain a RETURN statement, presumably controlled by another conditional statement.

It is perhaps worth stressing at this point that GOTO statements in all programming languages should be avoided if possible because it is very easy to produce unreadable, unreliable or wrong code by indiscriminate use of GOTO. In many modern languages it is possible to program entirely without using GOTO, and indeed the statement does not exist in Modula-2, the successor to Pascal. In languages such as FORTRAN 77 and REDUCE it is not possible to avoid GOTO entirely, mainly because in both languages the constructs necessary for breaking out of a loop are missing, and in other languages GOTO used with discretion can aid efficiency and avoid what would otherwise be very convoluted code.

3.7.4 Identifier scope: global versus local

The *scope* of an identifier in a program is the section of the program over which the identifier has a particular meaning; if the scope is restricted to some section then the scope is said to be *local* to that section, otherwise it is said to be *global.* These concepts should be familiar to users of ALGOL-based languages such as Pascal, but may not be familiar to users of FORTRAN and some versions of BASIC. In FORTRAN 77 variable identifiers are always local[6] to subprograms, and identifiers of subprograms and a few other objects are always global; the user has almost no control (with the exception of the EXTERNAL statement) over whether identifiers are local or global, so that identifier scope is rarely discussed in the context of FORTRAN. Similarly in "classical" BASIC all identifiers are global.

The algebraic mode of REDUCE is a language processor implemented as an "extension" of Standard Lisp (SLISP), and consequently it inherits most of the SLISP rules governing such matters as identifier scope. However, some features have been implemented using mechanisms different from those used in Lisp, with consequently slightly different scoping rules. The principal example of this is algebraic variable values. This makes the details of algebraic-mode variable scoping

[6]Variable *values* can be globalized using COMMON.

complicated, and impossible to describe fully without reference to symbolic mode (i.e. SLISP). Hence, you may find it useful to return to this section after having digested chapters 5 and 6. In particular, *global* and *local* are used as technical terms in Lisp with slightly more specific meanings than those described above.

By default, most identifiers in REDUCE act as if they are *global*, which means that a given identifier refers to precisely the same object wherever it appears within a program. For small programs this is no problem, and indeed it can be quite useful for keeping track of what is happening when code is being developed or debugged. However, for all identifiers to be global in a large program would be a disaster, because it is very hard to keep track of all those used and what they mean, especially in a program developed over a long period of time by many people. REDUCE itself is just such a large program, and provides little protection for many of the "internal" objects with which the user is not supposed to interfere.

Some assistance in avoiding identifier clashes is provided by the "Standard Lisp Cross Reference" facility called `CREF`, which is described in §18.3 of the REDUCE manual (Hearn 1987). It is supplied in the source file RCREF, which can be used if it is not available in compiled form – see also §8.1. `CREF` provides a summary of the identifier usage within a set of REDUCE source files. However, it is intended mainly for the symbolic-mode programmer, and will represent algebraic-mode objects in terms of their internal symbolic-mode forms.

The best way to avoid interference with global variable values, in situations in which it can be used, is the following. Within a block (and nowhere else) variables can be declared to be local. A local variable declaration can also restrict the data type of the variable, but it need not – REDUCE is not a strongly typed language. If a variable is local to some construct, then it is effectively a new object within that construct, which is used instead of, and without affecting, any existing object with the same identifier. Hence, it does not matter whether or not that identifier has been used elsewhere[7]. We have already met this idea: the control variables defined in for-loops and the "place holders" defined in let-rules with a `FOR ALL` qualifier are all local variables, local to the construct in which they are defined.

However, some identifiers are protected by having been declared `GLOBAL` or flagged `RESERVED` (in symbolic mode). Variables that have

[7]Unless the code within the block attempts to use *both* the local and the global values, which can happen in rather opaque ways because of the way that some REDUCE algebraic-mode facilities such as let-rules operate: see §§4.3, 4.7.1 and 4.7.3.

been declared GLOBAL in this way *cannot* be used as local variables. Variables that are used without being declared are actually "fluid" in Standard Lisp terminology, and can be declared and used elsewhere as local variables as described in the next section. This actually makes them either "fluid" or "local" as Lisp variables , depending on whether the code is compiled or interpreted and whether or not a symbolic-mode FLUID declaration is also used, as described more fully in §§4.3 and 4.7.1, but we shall usually refer to both as local.

3.7.5 Declaring local variables in a block

At the beginning of a block or compound statement, i.e. *immediately* after the BEGIN, lists of variables can be declared to be either SCALAR, INTEGER or REAL, as illustrated below. Variables declared to be SCALAR can take arbitrary algebraic values, including arbitrary numerical values, whereas variables declared to be INTEGER or REAL are intended to take only the specified types of values. However, as far as we have been able to ascertain, in the current version of REDUCE the only distinction among these declarations is that variables declared to be INTEGER are initialized to integer zero (0) whereas other local variables are initialized to polynomial zero (`nil` – see §5.3). The appropriate declaration should still be used to make the intention of the code clearer to a human reader, and for forward compatibility with future versions of REDUCE, which may take advantage of restrictions on the types of variables to generate more efficient code than is possible for the general scalar type. (Indeed, this may already be the case in some situations that we have not discovered.)

Local variables unfortunately have the potential disadvantage that when they are declared they are unavoidably initialized (to nil, which is a consequence of the way that local variables are implemented as Lisp "prog" variables (see §6.6) – variables declared integer are then explicitly reset to 0). Hence local variables are never unbound and always have a value, which by default is zero (or nil), so that local variables can never be kernels, and so can never be solved or substituted for. We will return to solutions of this problem in §§4.4, 4.7.3. Nevertheless, for many purposes it does not matter, and *variables should be declared to be local whenever possible.*

As an illustration, we return to our old task of converting a list to a matrix, and produce a version that does not interfere with its global environment either by using I as a *global* variable or by changing the value of THELIST:

```
THELIST := {A, B, C}$
MATRIX M(1,3);
% Either:
BEGIN INTEGER I;  I := 1;
   FOR EACH EL IN THELIST DO << M(1,I) := EL; I:=I+1 >>;
   RETURN M
END;

mat(1,1) := a
mat(1,2) := b
mat(1,3) := c

% or:
BEGIN SCALAR LOCLIST;  LOCLIST := THELIST;
   FOR I := 1 : 3 DO <<
      M(1,I) := FIRST LOCLIST;
      LOCLIST := REST LOCLIST >>;
   RETURN M
END;

mat(1,1) := a
mat(1,2) := b
mat(1,3) := c
```

In both cases the matrix `M` is accessed as a global variable within the block. A local matrix would be more appropriate here, but that is not so easy to arrange; we return to this problem in §9.2.

3.8 `WRITE` and `XREAD`

Only top-level statements, not those embedded in any kind of control or block structure, can produce output automatically if terminated by ";". However, the `WRITE` statement produces output as its main purpose, regardless of where in a program it appears and regardless of how it is terminated. The syntax is

`WRITE` *stmt1*, *stmt2*, ...

When executed, it starts a new output line, executes the statements *stmt1*, *stmt2*, ... (which must not be individually terminated) in sequence and outputs their values as a continuous stream, using as many output lines as REDUCE considers necessary, and with no formatting or automatic spacing whatever.

A string constant, which is a fixed sequence of arbitrary characters enclosed in double quotes, is a valid constant in REDUCE. One of the few useful things that can be done with such constants is to include them in `WRITE` statements in order to provide spacing and annotation, for example:

```
H := "hello"$
WRITE H, " world";

hello world
```

As with "top-level" output, the treatment of assignments by **WRITE** is anomalous in that they are output as complete assignment statements and not just as their values. Therefore, the following two statements produce the same output, although not necessarily the same effect since the second also *performs* the assignment, and if the environment has changed since a value was assigned to **H** then only the second statement will change the current value of **H**:

```
WRITE "H := ", H;

H := hello

WRITE H := H;

h := hello
```

Notice also that the case of letters in character constants is preserved. A **WRITE** with a null argument, such as an empty string, will output a blank line, but beware that a **WRITE** with no argument will be interpreted as a variable named **WRITE**, thus[8]:

```
WRITE "";

[ blank line output here ]

WRITE;

write
```

There is no "read" statement described in the REDUCE manual, which may seem strange to programmers familiar with other languages. However, with REDUCE there is rarely much occasion to use a read statement, and never any necessity. Input to REDUCE is normally provided either by explicitly entering assignments, or by invoking operators or procedures with arguments that provide the input. However, there may be occasions when it is convenient to be able to request input, and REDUCE itself clearly has an input reading capability. The procedure that REDUCE uses to input algebraic expressions is called **XREAD** (so called to avoid conflict with a standard Lisp input function called **READ**), and although not advertised **XREAD** is declared so as

---

[8]The variable **WRITE** could, and probably should, be defined as a synonym for "**WRITE ""**". There is also a need for a version of **WRITE** (called say **WRITEON**) that does not output an initial newline.

to be directly callable by the user from algebraic mode. It reads an expression in the form of a complete *terminated* REDUCE statement and returns the value of the statement, if any, after executing it. For example, one could implement a simple algebraic calculator like this:

```
WHILE (X := XREAD()) NEQ STOP DO WRITE X;
```

This simple loop behaves just like a restricted version of REDUCE itself, but without printing any prompts, until the identifier `STOP` is entered. However, `XREAD` reads from the current input stream, so that if it is executed from an input file then it will read the next statement from that file, which is unlikely to be what is required. To avoid this, `XREAD` needs to be embedded in a (symbolic) procedure that controls the I/O streams, and an example that always prompts to the screen and reads from the keyboard is given in the Appendix, §C. (Alternatively, the underlying Lisp read functions could be used to read individual characters, and for even more control it is usually possible to call the operating system directly, but this all requires programming in symbolic mode.)

3.9 Arrays

The remaining type of structured variable available in REDUCE is the array. Arrays were essential in earlier versions of REDUCE, but the introduction of algebraic-mode lists in REDUCE 3.3 has made arrays almost redundant. However, if a structured object is to be accessed by indexing then an array is almost certainly the most efficient structure to use both in terms of storage space and access time, because it is implemented internally as a block of contiguous memory (a *vector* in Lisp nomenclature – see §5.2.2), whereas a matrix is implemented as a list of lists.

Arrays in REDUCE are much like arrays in any other language, and can be multi-dimensional. They are declared with a list of upper index bounds for each dimension, the lower bound of which is always 0. Array elements can have any type of expression assigned to them, and are accessed in the usual way like matrix elements, except that they are not restricted to 2 indices. The operator `LENGTH` applied to an array identifier returns a list of the number of elements in each dimension of the array (i.e. its declared upper bound plus one).

The following example constructs and outputs an array of the first 20 Fibonacci numbers (this example is illustrative rather than efficient):

```
ARRAY F(20);  F(1) := F(2) := 1;
```

```
f(1) := f(2) := 1
```

```
FOR I := 3 : 20 DO WRITE F(I) := F(I-1) + F(I-2);
```

```
f(3) := 2
f(4) := 3
...
f(19) := 4181
f(20) := 6765
```

Arrays are global in a more fundamental sense than are matrix variables, because they are implemented in terms of Lisp vectors which themselves are global objects. Like matrix elements, array elements have the initial value 0 and so can never be kernels[9]. But, unlike matrices, there are no operations at all defined on arrays as composite objects. An array name on its own cannot be used for any purpose other than to clear the entire array including its values[10]. An attempt to clear a single array element generates an error unless the array element is a kernel, in which case its value is cleared; to reset an array element to zero simply assign zero to it.

3.10 Batch use

Most computers provide a mechanism to allow programs to be run non-interactively, and on many mainframe computers such program runs would be referred to as "batch jobs", because batches of non-interactive jobs are run when the operator, or more likely these days the operating system, chooses, and the user has no control whatever over the program while it is run. On a personal microcomputer, and even on many much larger modern machines, this concept is a little artificial, but it can be useful to run REDUCE with a minimum of intervention. Input must come from a file, and normally output would go to a file also.

Batch use makes sense for a "production run" of some large calculation involving a program that has been thoroughly tested interactively, or for tasks like compiling large REDUCE programs (or re-compiling parts of REDUCE itself). The only real difference between batch and interactive use of REDUCE lies in the precise form of invocation, which

[9]Arrays can be used in symbolic mode (see chapter 5 et seq.) *exactly* as in algebraic mode, the only difference being that arrays declared in symbolic mode have their elements initialized to `nil` rather than 0.

[10]This and the clearing of array elements were slightly different in previous versions; see also the Appendix, §D.

is highly system-dependent. Usually the input and output files will be specified in the same command that sets REDUCE running in batch mode, and REDUCE will then treat the input *exactly* as if it were coming from the keyboard. The input file should therefore end with "`BYE;`" just like an interactive session, otherwise the batch job may not terminate properly. The usual "`;END;`" after the "`BYE;`" will do no harm, but if placed before the "`BYE;`" it will switch into the underlying Lisp mode as remarked in §2.3, so that any subsequent REDUCE statements will be misinterpreted.

[This overloading of the `END` statement is probably a mis-feature of REDUCE; the reason for it is that REDUCE is actually started by a Lisp function called `BEGIN`, as described in chapter 5. This is hidden from the casual user, but it can be used by typing

```
(BEGIN)
```

to re-enter REDUCE from the underlying Lisp.]

As an example of batch use, REDUCE could be run in "background" mode on a UNIX system by giving a command of the form

`reduce <` *InputFile* `>` *OutputFile* `&`

Many microcomputers provide command line interpreters (CLIs) that are like a UNIX shell, and in which a command similar to that above (but without the `&`) will work. MS-DOS/PC-DOS systems provide such a CLI by default, and for a machine like the Atari ST there are many proprietary and public-domain CLIs available – see the Appendix, §A. (UNIX proper is case-sensitive and tends to use lower case preferentially – other systems may not care.)

Alternatively, the Lisp supporting REDUCE may itself provide a batch mode, e.g. Cambridge Lisp on the Atari would accept a parameter string of the form

`IMAGE` *ReduceImage* `FROM` *InputFile* `TO` *OutputFile*

where the `IMAGE` parameter is whatever is normally used to run REDUCE interactively.

3.11 Exercises

1. Use a `FOR`-loop to construct the 10×10 unit (or identity) matrix. Use another `FOR`-loop to print it out in the usual tabular form. Try [but not too hard, because you are unlikely to succeed generally] to write some code to do this in a fairly general way, perhaps by turning each row into a list before writing it.

2. Use REDUCE to generate, or otherwise construct, a list of objects to be used as polynomial coefficients, and then turn it into a polynomial using a `FOREACH`-loop.

3. The first and third terms of an arithmetic series are a and b respectively. The sum of the first n terms of this series is denoted by S_n. By defining operators to represent the rth term and the sum of the first n terms of an arithmetic series, or otherwise, find S_4 in terms of a and b. Given that S_4, S_5 and S_7 are consecutive terms of a geometric series, show that $7a^2 = 13b^2$.

4. Experiment with the generation of an approximation to $\sin x$ as a repeated product of linear factors constructed from its known zeros in the neighbourhood of $x = 0$, and compare it with its Taylor expansion about $x = 0$ truncated at the same degree. Use `FOR ... PRODUCT` and `FOR ... SUM` loops respectively to generate these. Is there a "best" set of zeros to use in the repeated product construction? [The repeated-product approximation will clearly have coefficients involving π, so you will need to do the comparison numerically – if you have access to REDUCE graphics then this is an ideal application.]

5. Write and test a general "program" to count the number of digits in a non-negative integer, supplied as the value of a global variable `N`. The program can be input and executed from a file after a value has been assigned to `N`. [However, after having read the next chapter it would be more elegant to re-implement the code as a procedure that uses the variable `N` as its argument.]

6. Write and test a general "program" to output all the prime numbers less than or equal to the positive integer value of a global variable `N`. Give some consideration to efficiency. Note that you can test the divisibility of two integers by simply dividing them and then testing the denominator, or by calling the `GCD` function. [See the comments on REDUCE "programs" in the previous question.]

7. Using the `MATEIGEN` operator, or otherwise, find the eigenvalues and eigenvectors of the matrix

$$A = \begin{pmatrix} -2 & 5 & 4 \\ 5 & 7 & 5 \\ 4 & 5 & -2 \end{pmatrix}.$$

Construct from the normalized eigenvectors of A a real orthogonal matrix X which reduces A to diagonal form via a transformation of the form X^TAX, where X^T denotes the transpose of X, calculate *explicitly* the diagonal matrix to which A is reduced, and verify *explicitly* that X is orthogonal, i.e. that $X^TX = XX^T = I$, where I is the unit or identity matrix.

8. Define an operator whose value for a positive integer argument n is the nth Fibonacci number f_n computed recursively from the formulae

$$f_1 = 1, \quad f_2 = 1, \quad f_n = f_{n-1} + f_{n-2} \text{ for integer } n > 2.$$

9. Define a Kronecker delta operator `DELTA(R,S)` with the following property:

$$\delta_{r,s} = \begin{cases} 1 & \text{if } r = s \\ 0 & \text{otherwise,} \end{cases}$$

test it and keep it for later use.

Try to do the same thing for the Dirac delta function `DELTA(X)` with the following property:

$$\int_a^b \delta(x - x_0) f(x) dx = \begin{cases} f(x_0) & \text{if } a < x_0 < b \\ 0 & \text{otherwise,} \end{cases}$$

for simplicity ignoring the cases $a = x_0$ and $b = x_0$.

10. Define a factorial operator that for non-negative integer arguments remembers all the values that it computes, automatically re-uses them, and uses the largest factorial computed so far to compute a new factorial. [It is possible to do this quite succinctly without using any extra operators, procedures or global variables.]

4

Algebraic procedures, operators and algorithms

After looking at algebraic procedures and at further details of operators, we go on to use REDUCE in some larger and more complete examples that involve some simple algebraic algorithms (for an introduction to algebraic algorithms in general see Davenport *et al.* 1988).

4.1 Procedures

Procedures are an essential feature of REDUCE, and it is procedures that ultimately do almost everything. However, we have left them until last so as not to distract from the other less conventional features of REDUCE. There is nothing that can be done with an algebraic procedure that cannot be done very similarly with an operator, although there may be efficiency reasons to use procedures and in many cases procedures seem more appropriate. We will compare operators and procedures in §4.3.

Like other statements, a REDUCE procedure call may or may not return an algebraic value, so that what would be called respectively functions and procedures in Pascal, or functions and subroutines in FORTRAN, are all procedures in REDUCE.

The syntax of a procedure definition is

PROCEDURE *name*(*arg1*, *arg2*, ...);
body-statement;

As usual, if there is only one or no dummy argument then the parentheses can be omitted. The body-statement is syntactically a single statement providing the body of the procedure, i.e. the code that is executed when the procedure is called; the procedure returns the value of this statement (which may apparently be nothing in algebraic mode). The procedure arguments are local to the procedure; they behave like variables that have been declared to be local within a compound statement, but which are initialized to the values with which the procedure is called rather than simply zero. Hence, the arguments can be used freely and safely as variables within the procedure body (but see the proviso below). Procedures can be defined at "top level" only (unlike in Pascal) and hence procedure definitions always have global scope (see §3.7.4).

A procedure is called (or executed) by simply giving its name followed by either a parenthesized argument list, a single argument or empty parentheses `()`[1]. If a procedure name is not followed by something that REDUCE can interpret as an argument then the procedure call will be misinterpreted as a variable reference. There is no `CALL` statement like that required in FORTRAN to call a subroutine. Procedures behave just like operators or the functions that are built into REDUCE, and procedures *should not* also be declared as algebraic operators. Procedures can be deleted or cleared by listing just their identifiers (without arguments) in a `CLEAR` statement.

The simplest form of procedure has just a simple statement as its body. (Pascal programmers note: a procedure body need not be a block.) For example:

```
% Define a procedure:
procedure cosec x; 1/sin x;

cosec

% and use it:
COSEC(PI/4);

    2
---------
 sqrt(2)
```

We are now in a position to provide a rudimentary definite integration facility. However, we will only consider the rather trivial problem of definitely integrating a function that is indefinitely integrable between finite limits within which the integrand has no singularities. Despite being trivial this capability is still quite useful. Removing the assumptions made above would provide an interesting research project, although some other computer algebra systems already provide quite sophisticated facilities for definite integration.

The best way to write a simple definite integration procedure is to use a group statement for the body together with the fact that procedure arguments are local, thus:

---

[1]Pascal does not allow empty parentheses in this context.

```
procedure DefInt(f, x, lower, upper);
   <<f := int(f, x); sub(x=upper, f) - sub(x=lower, f)>>$

DEFINT(SIN X, X, 0, PI);

2
```

There are at least two other ways to write this procedure, the first of which is grossly inefficient because it performs exactly the same integration twice, and this kind of inefficiency should always be avoided; the second is slightly less elegant and *might* be slightly slower:

```
procedure DefInt!-No(f, x, lower, upper);
   sub(x=upper, int(f, x)) - sub(x=lower, int(f, x));

procedure DefInt!-Maybe(f, x, lower, upper);
   begin scalar IndInt;
      IndInt := int(f, x);
      return sub(x=upper,IndInt) - sub(x=lower,IndInt)
   end;
```

Note that in the above examples **f** is an *expression* rather than the *name* of a function, because using an expression is more general. However, it is perfectly possible to pass operator names etc. as arguments, as in other languages; here is a re-write of **procedure DefInt** using this model, which is perhaps less elegant because of the need to invent the unbound variable "**DefInt!-Var**", a point to which we will return later.

```
procedure DefIntOp(f, lower, upper);
   <<f := int(f(DefInt!-Var), DefInt!-Var);
     sub(DefInt!-Var=upper, f) - sub(DefInt!-Var=lower, f)
   >>$

DEFINTOP(SIN, 0, PI);

2
```

[The above example procedures **DefInt** and **DefIntOp** both assign to **f**, which is a formal procedure argument. Rayna (1987, p. 183) strongly recommends against doing this on the grounds that it may cause problems, and recommends instead the programming style illustrated in our procedure **DefInt!-Maybe**. Assigning to formal arguments has never caused us any problems, and after examining the Lisp code that results from such procedure definitions in REDUCE 3.3 we do not see why it might be expected to cause problems. The problems may have been cured in REDUCE 3.3, or they may be specific to some implementations, or they may be something that we have missed, but in the event of difficulties Rayna's advice should be borne in mind.]

Statements within the body of a procedure never produce output automatically, so it makes no difference how they are terminated. The way to get output from a procedure is either to use a statement that has a value as the procedure body, to assign to global variables, or to use `WRITE` statements (although these mechanisms all produce different effects). However, if the *procedure body* (not the procedure statement itself) is terminated by a "`;`" then the procedure name is output when the procedure is defined; termination with a "`$`" prevents this (cf. respectively procedures `cosec` and `DefInt` above).

According to the manual, redefining an existing procedure should produce a warning message, which is particularly useful in case you inadvertently redefine an internal procedure. (This is quite possible, and provides a useful way of customizing the behaviour of REDUCE, as long as you know what you are doing and do it intentionally!) However, the generation of this warning is system-dependent, and in Cambridge REDUCE it is necessary to give (in advance) the instruction

```
LISP VERBOS 1;
```

in order to see the warnings. This also turns on messages about garbage collection. (The Cambridge Lisp function `VERBOS` takes a range of arguments; e.g. `VERBOS 3` will also generate messages about loading and unloading of REDUCE modules.)

4.1.1 Procedure type declarations

A procedure statement can be preceded by a type specifier, but the arguments then all have the same type as that specified for the procedure and cannot be independently specified. These are the types available:

- `ALGEBRAIC` is the default for a procedure defined in algebraic mode, and means that the procedure and its arguments have the general scalar type suitable for arbitrary algebraic expressions;
- `SYMBOLIC` (or `LISP`) defines a symbolic-mode procedure, to be considered in the following chapters;
- `INTEGER` or `REAL` means that the value (if any) of the procedure and its arguments are either integer or real respectively (although `REAL` is currently equivalent to `ALGEBRAIC`, cf. local variable declarations in §3.7.5).

The `INTEGER` declaration is not often used, but its advantage when it is appropriate is that it allows significantly more efficient code to be generated than is possible for a general algebraic procedure. An integer procedure *assumes* (without checking)[2] that its arguments are integers, and generates code based on that assumption. It *may* also check that its result is integer (apparently when it is not a direct consequence of the arguments being integers). Hence, there is a trade-off of efficiency against flexibility, which can be illustrated by considering a factorial procedure.

The following version returns the factorial of its argument `n` if it is a non-negative integer, or a suitable symbolic result otherwise.

```
procedure factorial n;
   if not fixp n then undefined
   else if n > 1 then for i := 2 : n product i
   else if n >= 0 then 1 else infinity;
```

(The variable identifiers `undefined` and `infinity` used here have no special significance to REDUCE, only to us.)

If we declared this to be an `integer procedure` in the interests of efficiency then it would be prevented from returning either of the symbolic results `undefined` or `infinity`, and the best that could be achieved in the event of a negative integer argument would be to return no algebraic value, as follows, although we provide a better solution to this problem in §4.2:

```
integer procedure factorial n;
   if fixp n then
      if n > 1 then for i := 2 : n product i
      else if n >= 0 then 1;
```

The factorial function applied to n is conventionally written $n!$, but there is no support in REDUCE for postfix operators (they are not very common in mathematics, anyway). The nearest we could easily get would be to name the procedure "`!!`" instead of "`factorial`", but it would still be a prefix operator.

4.1.2 Call by value; recursion versus iteration

REDUCE uses "call by value" to pass values to procedure arguments. This means that when the procedure is called, the values of the actual arguments with which it is called are *copied* into the dummy argument variables used when the procedure was defined. Therefore, the values of these argument variables can subsequently be changed without affecting the values with which the procedure was called, and the dummy

[2]The manual states that the type of the arguments and value of an integer procedure are checked, but our experience is that this is not at present generally the case, although this may change in future versions.

arguments can be used freely as local variables, as in our `DefInt` procedure above.

This contrasts strongly with the mechanism used in FORTRAN[3], which is "call by reference". With this mechanism, a dummy argument in a procedure corresponds to the same memory location as the actual argument used when the procedure is called, so that any changes made to the dummy argument also change the value of the actual argument. This mechanism provides another way of outputting information from a procedure, but it can also cause horrible and obscure errors. For example, in (some implementations of) FORTRAN it is quite easy to change the values of *explicit constants* by this means. Hence "call by reference" is not the default mechanism in modern languages, including REDUCE. However, it is possible to circumvent the call by value in a REDUCE procedure by using a let-rule to assign to a formal argument – see Rayna (1987), §4.8.1.2.

A procedure that is defined in terms of itself, i.e. that calls itself when it is executed, is said to be *recursive.* The purpose of recursive procedure calls is that each successive recursive call solves a simpler version of the problem, until finally the problem is trivial and can be solved without recursion, so the recursion terminates. The nature of the problem to be solved is represented by some data structure, which could be global, but more frequently is passed as an argument, in a progressively simpler form, to each recursive procedure call. This model of recursive procedure calls requires arguments to be passed by value; a simple example is the recursive factorial procedure illustrated below. Each recursive call must use new versions of any local variables, for which storage must therefore be allocated dynamically (i.e. as the program runs). Versions of FORTRAN before FORTRAN 90 allow all storage to be allocated statically and therefore do not allow recursion, although FORTRAN 90 does.

REDUCE uses call by value and dynamic allocation of memory for local variables, and allows recursive procedures. In fact in Lisp, which underlies REDUCE, recursion is the "pure" method of controlling program repetition. A recursive algorithm is usually less efficient than an iterative one, both in terms of memory usage and time, but if a problem is naturally recursive it may be much easier to implement a recursive solution, and hence much easier to produce a correct program. For example, the natural way to manipulate a multivariate polynomial, represented in REDUCE's standard recursive form, would be to use a recursive procedure, as will be discussed further in chapter 7 for

[3] In versions before FORTRAN 90, and it is still the default in FORTRAN 90.

the problem of factorization. Similarly, differentiating a complicated expression is obviously a recursive problem, as is searching for an occurrence of a particular variable within an expression, as illustrated by the modified version of the FREEOF predicate in the Appendix, §D. A recursive algorithm can often be converted into an iterative one later if necessary, and a compiler may do this automatically to some extent, thereby combining the clarity of the recursive definition with the efficiency of iterative evaluation.

The key to writing a recursive algorithm is to express the problem in terms of a simpler version of itself, and to have a trivial case (the so-called *base case*) that can be solved immediately. It is analogous to the process of mathematical induction. For example, the factorial of a positive integer n (a natural number) can be expressed as a repeated product

$$n! = \prod_{i=1}^{n} i$$

and we implemented this definition earlier. But it is easy to deduce that

$$n! = n \times (n-1)! \text{ for } n > 1 \quad \text{and} \quad 1! = 1,$$

and then to generalize this to

$$n! = n \times (n-1)! \text{ for } n > 0 \quad \text{and} \quad 0! = 1.$$

This leads to the following recursive implementation:

```
procedure factorial n;
   if not fixp n then undefined
   else if n > 1 then n*factorial(n-1)
   else if n >= 0 then 1 else infinity;
```

You will find that this implementation is significantly slower than the previous iterative version (although only obviously so for large arguments), and cannot compute such a large factorial in a given amount of memory. This procedure could be implemented more efficiently by making the recursive procedure a subsidiary one that does no error checking, along the lines of the example using an operator in §4.3.

4.2 Error exits from procedures and control structures

A well written program will always check that its input is valid, and if not it has to decide what to do. The programmer may be able to guess what a reasonable correction would be and to write the program so as to make it, preferably also outputting some kind of warning message

– this is what REDUCE itself often does. Alternatively, it may be necessary to abort execution of the program, but if the error is detected deep inside nested procedure calls, blocks or control structures then this is not easy to arrange. REDUCE provides a procedure called **RedErr** that causes a standard REDUCE error exit; after outputting the usual ***** it displays its argument, which should normally be a character string describing the error, exits from the whole nest of active REDUCE procedures, composite statements and control structures, restoring values of variables used locally if necessary, and returns to a new REDUCE input prompt. For example, a factorial procedure intended to accept only non-negative integer argument values could be written as follows:

```
integer procedure factorial n;
   if not fixp n then
      RedErr "Factorial: non-integer argument invalid"
   else if n > 1 then for i := 2 : n product i
   else if n >= 0 then 1
   else RedErr "Factorial: negative argument invalid";
```

Errors occurring during input from a file, including "errors" generated by calls of **RedErr**, will suspend input from the file and query the user unless the switch **ERRCONT** is turned **ON**; see also §2.3.

4.3 Procedures versus operators – compilation

The REDUCE manual blurs the distinction between operators and procedures in a way that we consider unhelpful, and when we use the term *procedure* we will mean specifically a set of instructions defined using a **PROCEDURE** statement. It is quite possible to make either a variable or an operator evaluate to any statement by using a let-rule. Making a variable evaluate to a statement, possibly a complicated block structure, provides a way of writing a "command" that takes no arguments and can be invoked without the need for empty parentheses, e.g. **SAME** in §2.10.1. Making an operator with arguments evaluate to a statement provides a way of writing the equivalent of a procedure whose arguments are passed "by name", i.e. by pattern matching, rather than by value, e.g. **QUADRATIC!-ROOT(A,B,C,PM)** in §2.11, and subsequent examples. (Call-by-name is an ALGOL 60 concept that is slightly more general than call-by-reference). See §4.10 of Rayna (1987) for further discussion.

Probably the main distinction between variables or operators that have evaluations defined using let-rules as described above and genuine procedures is that it is possible to *compile* procedures in most implementations of REDUCE, and certainly in REDUCE based on PSL or CULisp, both of which provide a compiler based on the Standard Lisp Compiler of Griss and Hearn (1981). This is because REDUCE procedures are converted by the parser into Lisp functions (using the internal Lisp forms of REDUCE's algebraic functions), which are all that the Lisp compiler actually compiles. (Similarly, it is possible to produce a fast-loading file of REDUCE code in most implementations[4], in which the procedures are compiled and the rest of the code is "partially digested", by using the `FASLOUT` command – see §18.2 of the REDUCE manual (Hearn 1987)).

Depending on what they do, procedures[5] can execute very much faster when compiled than interpreted. Hence, for serious tasks, it is usually worthwhile to consciously aim to use procedures rather than operators etc., and to compile them. The easiest way to use the compiler (in Lisps such as PSL and CULisp that have a built-in compiler) is to make the switch setting `ON COMP`, after which all procedures will be automatically compiled as they are input. (It is usually also possible to compile procedures after they have been defined by calling the compiler directly in symbolic mode.)

The other distinctions between procedures and operators are:-

- procedures cannot be defined within other procedures, whereas operators can be defined anywhere;
- operators are much more general, in that they can remain symbolic or can be conditionally evaluated, which cannot be done for procedures using only algebraic mode.

As an example of the greater flexibility of operators, and of how an operator can usefully be defined partially in terms of a procedure, thereby getting the best of all possibilities, here is yet another version of the factorial function:

```
operator factorial;
for all n such that fixp n and n >= 0 let
   factorial n = ifactorial n;
integer procedure ifactorial n;
   if n >= 2 then for i := 2 : n product i else 1;

% Examples of use:
FACTORIAL 5;

120
```

[4]But not in the current Atari implementation.

[5]Although this applies less to algebraic than to symbolic procedures.

```
FACTORIAL 1;

1

FACTORIAL 0;

1

FACTORIAL (-1);

factorial(-1)

FACTORIAL PI;

factorial(pi)
```

There should be a considerable advantage in having the for-loop in `procedure ifactorial` compiled if its argument is large. (In fact, there is a procedure very much like `ifactorial` already defined internally in REDUCE in the MATHLIB source file.)

There is one potential catch to compiling procedures, which is that the semantics (i.e. the meaning) of compiled Lisp can be slightly different from that of interpreted Lisp. In particular, in many implementations the compiler will discard references to variables by name in favour of pointers to their values if they are local, i.e. procedure arguments and variables declared to be local in compound statements, which have not also been explicitly declared FLUID. Usually this does not matter, but occasionally it does matter for reasons that are not immediately obvious but are due to the way that the parser converts algebraic-mode REDUCE code into Lisp. In particular, it seems to matter for let-rules used inside procedures. However, symbolic mode provides the "FLUID" declaration to tell the compiler to keep the names of specified variables, which are given as a quoted Lisp list of identifiers thus:

```
FLUID '(X Y Z);
```

We will illustrate this later in a real example in §4.13 – see also §3.7.4, §4.7.1 and chapter 6.

4.4 Local and pseudo-local identifiers

This section reiterates some of the remarks made already in §3.7.4, but now specifically in the context of procedures; some aspects that require reference to Lisp (i.e. symbolic mode) are deferred until §4.7. We have seen the advantages of using local variables, but we have also seen firstly that local variables are always bound and so are never kernels, secondly that some objects, such as procedures, operators, arrays, and to some extent matrices, cannot be local, and thirdly that little protection is given against accidental re-use of global identifiers.

In general, the only purely algebraic-mode solution available at present is to give local variables very obscure names. Global identifiers in REDUCE itself[6] often contain special characters such as *, #, :, all of which must be escaped (with !) when used. Hence you should avoid including these special characters in your own identifiers. However, beware that there are many REDUCE internal procedures whose names are not especially obscure.

One way of creating obscure identifiers is to use very long names, with parts that are very specific to your particular program. For example, they might include the name of the procedure in which they are being used, as in `procedure DefIntOp` earlier, or your initials, or your car licence plate number etc. This technique is admittedly very inelegant, but it is necessitated by limitations in Standard Lisp[7].

Returning to the second drawback mentioned above, we will show in chapter 9 how to treat matrices as genuinely local objects, and we anticipate that similar techniques could be applied to arrays and operators (though not to procedures) and other global objects. However, local arrays, matrices etc. can always be faked by giving them obscure names and making sure that they are completely cleared at the end of the segment of code in which they are used, which will often be a procedure body. This is illustrated at the end of the next section for a pseudo-local array. Of course, it is perfectly reasonable that one of the purposes of a procedure or other code segment might be to *create* some global object. In that case, one would not want to use an unduly obscure name for it, and a neat approach is to make the identifier to be used for the object an argument to the procedure. This is the topic of the next section.

The problem that local variables are not kernels has a more elegant solution, which avoids the risk of using variables not declared to be local, but it involves a very minor excursion into symbolic mode, so we defer it until §4.7.3 together with other "advanced topics".

---

[6]This includes `FLUID` variables which the user is not expected to access.

[7]Other Lisps have facilities variously called "packages", "flavours", "closures" etc. to avoid this problem, which is a classic of Lisp design! A Lisp family tree is presented in §5.2.1.

4.5 Passing objects to procedures

Passing scalar values to procedures is straightforward; we are concerned here with passing other objects. Algebraic-mode lists can be freely passed to and from procedures just as scalars can. The problem is passing global objects: it is possible to pass (by value) the *name* of a global object as long as it does not automatically evaluate, which works only for names of operators and procedures. A matrix identifier automatically evaluates to the whole matrix, and an array identifier used alone generates the error message[8] "`Array arithmetic not defined`". However, it is still straightforward to pass an identifier that is to be declared in the procedure to be a matrix or array identifier.

The following procedure illustrates this by generating the Hessian matrix of a (generally multivariate) function f. The Hessian matrix of a function is its matrix of second partial derivatives; its precise definition should be clear from the following code. The names of the variables on which f depends are passed as a list to the dummy argument `vars` and the name to be given to the global Hessian matrix is passed as the dummy argument `name`. The procedure also illustrates the use of a pseudo-local array as a more convenient way of accessing the variables internally than the list in which they are passed to the procedure. (This could have been handled differently by manipulating lists, and perhaps for purposes other than illustration this would be the best approach.)

```
procedure Hessian(f, vars, name);
   % vars is a list of the variables
   % on which the expression f depends
   % Call the (global) Hessian matrix name
   begin integer n;  n := length vars;
      matrix name(n, n);
      array Hessian!-Variable!-Array (n);
      for i := 1 : n do <<
         Hessian!-Variable!-Array i := first vars;
         vars := rest vars >>;
      for i := 1 : n do for j := 1 : n do
         name(i,j) := df(f, Hessian!-Variable!-Array i,
                            Hessian!-Variable!-Array j);
      clear Hessian!-Variable!-Array
   end$

% Try it:
Hessian(x**2+y**2, {x,y}, h);
```

[8]New in REDUCE 3.3; in previous versions an array identifier could also be used as a variable identifier, and hence effectively arrays could be "passed by name" to procedures, but see the Appendix, §D, for a way to permit this again.

```
H := H;
```

```
h(1,1) := 2
h(1,2) := 0
h(2,1) := 0
h(2,2) := 2
```

4.6 Debugging procedures and control structures

When developing procedures or other kinds of structures such as block or control structures in algebraic mode, and something goes wrong, there is often nothing to indicate *what* has gone wrong. REDUCE detects syntax and type-mismatch errors, but cannot of course detect errors in program logic. Programs should always be developed in small well-defined pieces, and even if a program is written all at once it should be tested in small pieces. One aid to debugging is to insert `WRITE` statements at various places. Remember that the arguments of a `WRITE` statement can be statements, so one can simply insert `WRITE` in front of chosen statements. This can be done on the line above *as long as* `WRITE` *itself has no terminator*, and then once the code is working the `WRITE`s can be simply deleted or commented out, thus:

```
%%write "var = ",
   if SomeCondition then var := Something
      else var := Somethingelse;
```

An alternative is to comment out the "structuring", e.g. comment out a `PROCEDURE` statement, set up the procedure argument values by assigning to them as global variables, and then run the code (e.g. by re-inputting from a file). If necessary, also comment out group delimiters (`<<` and `>>`), and block delimiters (`BEGIN` and `END`) and `RETURN`s together with any local variable declarations, `GOTO` statements and labels. The value of each statement will then be automatically output (assuming that it has been terminated with ";", which is one good reason for using this terminator inside composite statements). This should make it clear where the code is failing and perhaps why. A half-way step to this technique is to comment out just the local variable declarations in a block, so that the variables become global and can be examined after a faulty block has failed.

As a last resort, the symbolic-mode debugging techniques described in chapter 6 can be applied, and indeed if the problem is caused by a bug in REDUCE itself or in the Lisp system then this will probably be essential. In particular, turning on the switch `BACKTRACE` will produce a Lisp backtrace after any error detected by the underlying Lisp, which whilst often being rather long can provide useful clues, especially once one understands the internal operation of REDUCE as described in subsequent chapters.

4.7 Advanced aspects of identifier scope

This section can be skipped by the beginner. It is included at this point in order to complete our discussion of algebraic procedures, although it involves references to symbolic mode, which is not introduced formally until chapter 6. Further examples of using symbolic mode from algebraic mode are given in chapter 9.

4.7.1 Global and fluid variables

Global and fluid are Standard Lisp technical terms that were briefly introduced in §3.7.4 and are described more fully in chapter 6. Also, `GLOBAL` and `FLUID` are symbolic-mode "declarations" (i.e. Lisp functions), each of which expects a Lisp list of identifiers as its single argument. They are not intended to be used in algebraic mode but if, as is usually the case, their arguments are quoted lists then they are in fact directly accessible from algebraic mode. In interpreted code all variables are regarded as fluid by default, whereas in compiled code they are only fluid if explicitly declared so.

Attempting to re-use an identifier, usually accidentally, can give rise to two problems. The easier one is that some identifiers, in particular those which have been declared `GLOBAL`, cannot be used as local variables, either by declaring them within a block or by using them as procedure dummy arguments, and any attempt to do so gives rise to an error message. However, such `GLOBAL` identifiers can still be used as algebraic-mode global variables.

The more difficult problem is that a local variable which is considered *fluid* may elsewhere be the name of a procedure or (fluid) variable, which may be called or used by the procedure in which the identifier is being used as a local variable. For example, `MAXDEG` is the name of an internal procedure used in REDUCE's polynomial manipulation routines; if `MAXDEG` is used as a fluid variable in a procedure `test` then `MAXDEG`'s definition is temporarily suspended (stacked), so that if `test` requires a polynomial operation which happens to need the internal REDUCE procedure `MAXDEG` then `test` will find the local variable value instead of the required procedure definition, and therefore give rise to a rather baffling error. In *compiled* functions[9] this problem does not

---

[9]This behaviour depends on the compiler, and thus may be implementation-dependent; what we describe is the usual behaviour,

affect procedure argument names or local variables unless they have been explicitly declared FLUID.

4.7.2 Variable scope for algebraic procedures

Consider the following three algebraic-mode statements:

```
x := a$
procedure test;  begin scalar x; x := b; test1() end$
procedure test1; write "x = ", x$
```

The global variable x has been assigned the global value a, but x is also a local variable in procedure test, in which it is assigned a local value b. Procedure test1 accesses x as a global variable, and when called from top level would be expected to access the global value a of x, which it does:

```
TEST1();

x = a
```

But when procedure test1 is called from within procedure test, will it access the global value a or the local value b of the variable x? The answer is that it will always access the global value x:

```
TEST();

x = a
```

This contrasts with what can be done in symbolic mode, in Lisp itself, and in languages like Pascal (in which procedures can be defined locally). Exactly the same happens if procedure test1 is replaced by an equivalently defined operator.

This means that the only way using purely algebraic mode to pass *local* values between procedures is as arguments and *not* by using variables that are local to one procedure but global to the other. The reason is that local variables in algebraic procedures use Lisp values whereas the algebraic values of global variables are stored differently (see §5.3). Using this information, procedure test1 could be modified to use the Lisp value of x thus (where the operator SYMBOLIC could be replaced by LISP):

```
procedure test1; write "x = ", symbolic x$

TEST();

x = b
```

---

which conforms with the definitions in Marti *et al.* (1979).

This approach is quite safe. (However, if the code were compiled then the variable **x** would need to have been declared **FLUID** before **procedure test** was compiled – see §4.3 and also §3.7.4, §4.7.1 and chapter 6.) Alternatively, the variable **x** could be declared to be shared between algebraic and symbolic (Lisp) modes, which actually causes its algebraic value to be stored as its Lisp value:

```
SHARE X;

TEST();

x = b
```

However, this approach has the disadvantage that **x** can no longer be cleared using only algebraic mode[10]. Hence variables should not be casually declared to be **SHARE**d.

Note also that when compiling (in some implementations) it seems that variables that are **SHARE**d need to be explicitly declared as such in each file in which they are used, even if they have already been so declared elsewhere in the system, as for example when using the global shared variable **MULTIPLICITIES!***.

Of course, information can always be passed between procedures by using variables that are global to both of them, but this suffers from the insecurity common to all global variables.

4.7.3 Local unbound variables

Symbolic mode provides a function called **GENSYM**, which takes no arguments and returns an identifier that is guaranteed to be unused, and so must be unbound and can reliably be used as a kernel. This unbound identifier can be assigned to a normal local variable in order to access it. Here is a more elegant version of **procedure DefIntOp** written using this technique:

```
procedure DefIntOp(f, lower, upper);
   begin scalar Var;  Var := symbolic gensym();
      f := int(f(Var), Var);
      return sub(Var=upper, f) - sub(Var=lower, f)
   end;
```

We will illustrate this again in a larger example in §4.13.

The above technique will work only in situations where variables are (algebraically) evaluated, which does not include declarations, or

[10]The symbolic-mode statement "**REMFLAG('(X), 'SHARE)**" undoes the main effect of the declaration "**SHARE X**", and we have proposed an **UNSHARE** command that achieves this more elegantly and completely.

variables on the left of assignments or let-rules, etc. However, it is possible to use essentially the same trick to reliably construct local names for matrices and arrays as long as the *assignment* of the identifier returned by **gensym()**, as well as the actual call of **gensym()**, is made in symbolic mode. This is illustrated by the following more elegant version of the Hessian procedure introduced in §4.5.

```
procedure Hessian(f, vars);
   % vars is a list of the variables
   % on which the expression f depends
   % Returns the Hessian matrix of f wrt. vars
   begin scalar h, v;  integer n;
      symbolic << h := gensym(); v := gensym() >>;
      n := length vars;
      matrix h(n,n);  % The Hessian matrix
      array v(n);     % An array to hold the variables
      for i := 1 : n do <<  % List -> array
         v i := first vars; vars := rest vars >>;
      for i := 1 : n do for j := 1 : n do
         h(i,j) := df(f, v i, v j);
      clear v;
      return h
   end$

% Try it:
Hess := Hessian(x**2+y**2, {x,y});

hess(1,1) := 2
hess(1,2) := 0
hess(2,1) := 0
hess(2,2) := 2
```

In order to use this trick outside of a block, for example for interactive experimentation, it is necessary to declare any variables to which identifiers returned by **gensym()** are assigned to be **SHARE**d, which itself entails the minor problems referred to in the previous subsection.

4.8 Declaring general algebraic properties

4.8.1 Infix operators and procedures

Operators or procedures can be declared to be infix, as opposed to prefix. Mathematically, an infix operator is one that is placed between its two operands instead of in front of them; infix operators are necessarily binary, whereas prefix operators can take an arbitrary number of arguments. Declaring a REDUCE operator or procedure to be infix affects only the way it is input and output and not the way it is treated internally; prefix form can still be used input. The purpose of the infix declaration is to allow REDUCE programs to look more like conventional mathematics. For example, when working with vectors it might be useful to define infix operators to represent scalar and vector products, which can then be used in algebraic expressions together with +, -, etc. (This is done properly in the vector calculus package which will be discussed in chapter 8 and is written mostly in symbolic mode.) A separate operator declaration is not necessary for identifiers that are declared infix, e.g.

```
INFIX DOT, CROSS;
A DOT B;

a dot b

A CROSS B;

a cross b

A DOT B + A DOT B;

(a dot a + b) dot b
```

The last expression has clearly been parsed wrongly, which has happened because the default *precedence* of the operator DOT is wrong and has been put lower than +. The precedences of the built-in arithmetic operators are, in decreasing order, ** / * - +. To give our DOT operator multiplicative-type precedence we need to put it just before "-" as follows:

```
PRECEDENCE DOT,-;
A DOT B + A DOT B;

2*(a dot b)
```

The infix declaration also works for procedures, and if the declaration is made before the procedure definition then the procedure can be defined in infix form if desired, as illustrated in the following rather useless example:

```
INFIX CROSS1;
PROCEDURE A CROSS1 B; WRITE A, " X ", B$

A CROSS1 B;

a x b
```

4.8.2 Symmetric and antisymmetric operators

Operators of any type, prefix or infix, can be declared to be either symmetric or antisymmetric with respect to *all* their arguments. This means that the arguments are re-ordered into a standard order (which by default is system dependent), making appropriate sign changes to the value of an antisymmetric operator and simplifying it to zero if two arguments are equal. This could be useful for defining a symbolic determinant or alternating tensor, and is useful for our scalar and vector product operators:

```
SYMMETRIC DOT;  A DOT B - B DOT A;

0

ANTISYMMETRIC CROSS;  A CROSS A;

0

A CROSS B;

a cross b

B CROSS A;

 - (a cross b)
```

Additional properties of these operators would need to be defined using let-rules, e.g.

```
A DOT (B+C);

(b + c) dot a

FOR ALL A,B,C LET A DOT (B+C) = A DOT B + A DOT C;
A DOT (B+C);

a dot b + a dot c
```

4.8.3 Non-commuting operators

Operators can be declared to be non-commuting under the built-in multiplication operation *. This is done by the `NONCOM` command, and only works for operators, not for variables, although it is not necessary that the operators have any arguments. Operators declared to be non-commuting that differ either in their identifier or their argument list will not be commuted by REDUCE. The concept of "non-commuting procedures" does not make sense because procedures always evaluate to something, but the objects to which they evaluate could be non-commuting operators. A well known pair of non-commuting operators from quantum mechanics are the position and momentum operators, traditionally called q and p:

```
OPERATOR Q, P;  NONCOM Q, P;
```

```
Q()*P() - P()*Q();

 - (p()*q() - q()*p())

% But:
Q*P - P*Q;

0
```

The arguments of q and p in quantum mechanics are often left implicit, as above, and it would be nice to be able to regard them as non-commuting variables. Similarly, it would be nice to be able to represent non-commuting group elements as REDUCE variables, but this is not possible at present. The nearest that one can get is to avoid the () on input, thus:

```
LET Q=Q(), P=P();
Q*P-P*Q;

 - (p()*q() - q()*p())
```

but there is no way to avoid the () on output, because if the operators are replaced by variables then they immediately commute! It is easy to define a let-rule to commute non-commuting operators into a standard order with the introduction of a commutator. Quantum mechanists will recognize the following example:

```
LET Q()*P() = P()*Q() + I*HBAR;
Q*P;

p()*q() + hbar*i
```

An example of a more sophisticated commutator rule is given in the REDUCE manual (Hearn 1987), where the non-commutativity is of one operator with two different arguments.

4.8.4 Linear operators

Many operations are linear, e.g. differentiation and integration, and tortuous let-rules can be avoided by simply declaring a REDUCE operator to be linear, using the command `LINEAR`. However, REDUCE needs to know the variable with respect to functions of which the operator is linear, and this variable must always be the second argument of a REDUCE linear operator, since otherwise REDUCE has no way of distinguishing between variables and symbolic constants. Any remaining arguments are ignored as far as linearity is concerned. As an illustration, let us use REDUCE to remind us of the mathematical definition of a linear operator L operating on functions $f1$ and $f2$ of x

with $a1$ and $a2$ in the field of constants:

```
OPERATOR L;  LINEAR L;
OPERATOR F1, F2;
L(A1*F1(X) + A2*F2(X),X);
```

```
l(f1(x),x)*a1 + l(f2(x),x)*a2
```

The first argument of a REDUCE linear operator can depend on the second in any way, either explicitly as an operator or expression, or implicitly through a DEPEND declaration. This and the ability of a linear operator to carry extra arguments are illustrated by the following example of a symbolic definite integration operator:

```
DEPEND FF1, X;  DEPEND FF2, X;
OPERATOR DEFINTLIN;  LINEAR DEFINTLIN;
DEFINTLIN(A1*FF1 + A2*FF2, X, LO, HI);
```

```
defintlin(ff1,x,lo,hi)*a1 + defintlin(ff2,x,lo,hi)*a2
```

(Implicit dependence can be cleared with a NODEPEND declaration that is otherwise identical to the DEPEND declaration that it is intended to undo, as was explained at the end of §1.5.)

One final difference between operators and procedures is that procedures cannot be declared to be either symmetric, antisymmetric, non-commuting or linear (actually they can, but it has no effect!). If necessary, these properties can be declared for operators that are then made to evaluate to procedures using let-rules.

...

We have now covered almost all of the algebraic mode of REDUCE, and almost all of it that we intend to cover in this book. The rest of this chapter is devoted to some longer examples that are mainly procedural, and provide an introduction to some simple algebraic algorithms and their implementation in REDUCE 3.3. Note that none of the code presented is intended to be bullet-proof distribution-quality code, but rather it is intended to be illustrative. It is left as an exercise for the reader to generalize the examples, add more error checking etc.

4.9 Taylor expansion

It is perhaps surprising that there are no facilities for generating series expansions (binomial, univariate and multivariate Taylor, Laurent etc.) built into REDUCE, but simple versions of all these are easy to program. As an example, we will consider two versions of a Taylor expansion procedure; the other expansion problems are relegated to exercises. The correct techniques for summing (i.e. generating) series algebraically are similar to those that one would use numerically, and the most important point is not continually to re-compute the same value. Sequences and series often have a regular structure that allows each term to be generated efficiently from the previous term (or occasionally terms). This involves a trivial kind of recursion to build both the terms and the sum itself, as illustrated in the following straightforward Taylor procedure, which is programmed in the same style that one would (have to) use in conventional languages like FORTRAN.

```
procedure taylor(f, x, x0, n);
   % Return the Taylor expansion of f with respect to x
   % about x0 up to degree n
   begin scalar fact, series; % local variables
      fact := 1;              % factor multiplying deriv
      series := SUB(x=x0, f); % constant term of series
      for i := 1 : n do <<    % for each non-constant term
         f := df(f, x);       % recursively differentiate,
         fact := fact*(x-x0)/i;         % generate factor
         series := series + sub(x=x0, f)*fact % and series
         >>;
      return series
   end$
```

Here is a slightly more sophisticated version of the same procedure:

```
% For Cambridge REDUCE only,
% to turn on procedure re-definition messages:
lisp verbos 1;

procedure taylor(f, x, x0, n);
   begin scalar fact, deriv;
      fact := 1;  deriv := f;
      return sub(x=x0, f) +
         for i := 1 : n sum
            sub(x=x0, deriv := df(deriv, x)) *
            (fact := fact * (x-x0)/i)
   end$

*** taylor redefined
```

This version uses a more functional style of programming, which is not available in most conventional languages (apart from C, to some extent), i.e. it uses the values of the assignments as it makes them, and it sums the series using a `FOR ... SUM` loop. Because one cannot assume left-to-right evaluation of expressions, one cannot assume that the `SUB` to evaluate the constant term will be executed before the for-loop. Hence in this version of the procedure it is not reliable to use `f` internally to represent the derivative because it might have been differentiated n times before being used to evaluate the constant term, so the local variable **deriv** has been introduced and used rather than **f**. These two procedures – using whichever style you prefer – provide a model for most other series expansion problems.

We will now demonstrate that our (second) Taylor procedure works:

```
OFF ALLFAC; ON DIV, REVPRI;  % To improve appearance
% Expand tan x and arctan x about 0 to degree 7:
TANX := TAYLOR(TAN X, X, 0, 7);

              1   3     2   5     17    7
tanx := x + ---*x  + ----*x  + -----*x
              3         15       315

ATANX := TAYLOR(ATAN X, X, 0, 7);

                        1   3     1   5     1   7
atanx :=  - ( - x + ---*x  - ---*x  + ---*x )
                        3         5         7

% Check by substituting one into the other,
% but ignore expected error terms (of order 9):
LET x**9 = 0;
SUB(X=ATANX, TANX);

x

SUB(X=TANX, ATANX);

x
```

4.10 Levels of abstraction – how symbolic is symbolic?

In the same way that conventional languages do not support variables with no assigned value, so computer algebra systems only support symbolic quantities to some level of abstraction. For example, one might wish to work with indefinite finite sums of the form

$$\sum_{j=1}^{n} f(j)$$

with unspecified n, where $f(j)$ is some (specified or unspecified) function of j. The obvious transcription into REDUCE is

```
for j := 1 step 1 until n sum f(j);
```

but REDUCE will give an error unless **n** evaluates to an explicit integer constant, because REDUCE always tries to *execute* a for-loop when it encounters it, and it cannot loop an unspecified number of times.

This level of abstraction is beyond the capabilities *explicitly* offered by REDUCE. The simplest solution is to define a new operator to represent the finite sum, and perhaps define the value of the operator in the case that n has an explicit integer value by a suitable rule, thus:

```
operator finite!-sum, f;
for all f, n such that fixp n let finite!-sum(f, n) =
   for j := 1 step 1 until n sum f(j);
```

Computer algebra systems differ in the level of abstraction that they support; Maple has for some time had built-in support for symbolic finite sums plus code to evaluate some classes, and a similar user-contributed REDUCE package has recently been announced. An analogous problem occurs with symbolic n^{th} derivatives, which is our next topic.

4.11 Symbolic n^{th} derivatives and Leibniz' formula

Leibniz' formula gives the nth derivative of a product of two functions. It is normally used when n is symbolic (otherwise one may as well just differentiate explicitly n times) and is most useful when one of the factors is a polynomial and so eventually differentiates to zero. This presents two problems for its implementation in REDUCE.

The first problem is that the REDUCE differentiation operator `DF` cannot be used with a symbolic order of differentiation, so we have to define our own symbolic n^{th} derivative operator `DFN`, give it all the necessary properties that are predefined for `DF`, and make it evaluate to `DF` when n has an explicit integer value.

The second problem is to decompose a product into two factors, one of which is a power of the independent variable to be differentiated explicitly and the other is therefore to be differentiated symbolically. If this decomposition is not possible the `DFN` operator should be left symbolic, because it will not be possible to express the result as a known number of terms. We perform the decomposition by using a single qualified let-rule, whose condition relies on the fact that `DFN` has been declared `LINEAR` so that after simplification it will only be applied to products of functions of the independent variable, of which any monomial factor must be a polynomial with leading coefficient 1 – any other factor has leading coefficient equal to itself and hence not 1. Note that the let-rule used below, which involves explicitly only one multiplication, is enough to decompose a product of an arbitrary number of factors, and the `LINEAR` declaration ensures that these factors will all depend on the independent variable in some (explicit or implicit) way.

This example provides a good illustration of the combined use of operators and procedures – the symbolic part of the problem is all handled by operators and let-rules, whereas the part that requires actual computation is handled by a procedure, which could be compiled for speed. It also illustrates how one can handle symbolic problems that appear at first sight to be beyond the capabilities of REDUCE.

```
% Establish an n-th order derivative operator that allows
% n to be symbolic and applies Leibniz' rule to products:

operator dfn;  linear dfn;

for all y, x, n such that y freeof x let dfn(y, x, n) = 0;

for all y, x, n such that fixp n and n >= 0 let
   dfn(y, x, n) = df(y, x, n);

for all y, x, n let
   dfn(df(y,x), x, n) = dfn(y, x, n+1),
   df(dfn(y,x,n),  x) = dfn(y, x, n+1);

for all y, x, n, m such that fixp m and m >= 0 let
   df(dfn(y,x,n), x, m) = dfn(y, x, n+m);

% Check whether Leibniz can be applied, and if so
% decompose product so that FIRST argument to Leibniz
% is a power of x (because dfn is linear).
% Note: The following let-rule picks the appropriate
% factor out of an arbitrary product.

for all u, v, x, n such that lcof(u,x) eq 1 let
   dfn(u*v, x, n) = Leibniz(u, v, x, n);


procedure Leibniz (u, v, x, n);
   % Apply Leibniz' formula to df(u*v, x, n) with symbolic
   % n ASSUMING that u is polynomial in x. Hence n cannot
   % be used to create a stopping condition, but the
   % degree of u can be.
   begin scalar series, binom;  integer r;
      binom := 1;  r := 0;  series := u*dfn(v, x, n);
      while  (u := df(u,x)) neq 0  do  series := series +
         (binom := binom*n/(r:=r+1))*u*dfn(v, x, n:=n-1);
      return series
   end;
```

Here is a very simple example to demonstrate that DFN and LEIBNIZ work[11]:

---

[11]Testing this code showed up a problem in FREEOF – see the Appendix, §D, for further details.

```
DEPEND Y,X;
DFN(A*X*SIN(X)*Y, X, N);

dfn(sin(x)*y,x,n - 1)*a*n + dfn(sin(x)*y,x,n)*a*x
```

A typical use of Leibniz' formula is to derive a recurrence relation for derivatives from a differential equation, e.g. differentiate

$$(x^2 - 1)\frac{dy}{dx} + xy + 1 = 0$$

n times. This is now very easy using our DFN operator:

```
DEPEND Y, X;  FACTOR DF, DFN;  % to group terms
ODE := (X**2 - 1)*DF(Y,X) + X*Y + 1;

                    2
ode := df(y,x)*(x  - 1) + x*y + 1

DFN(ODE, X, N);

                  2                        2
dfn(y,x,n - 1)*n  + dfn(y,x,n + 1)*(x  - 1) +

         dfn(y,x,n)*x*(2*n + 1)
```

4.12 Constructing orthogonal polynomials

A set $\{p_k\}_{k=0}^{\infty}$ of polynomials where p_k has degree k are said to be *mutually orthogonal* on an interval $[a, b]$ with respect to a weight function $w(x)$ if

$$\langle p_i p_j \rangle = 0 \quad \text{if} \quad i \neq j$$

and they are said to be *normalized*, or *orthonormal*, if also

$$\langle p_i p_j \rangle = 1 \quad \text{if} \quad i = j,$$

where

$$\langle f \rangle \equiv \int_a^b w(x) f(x)\, dx.$$

We want to construct such a set of orthogonal polynomials, and perhaps also normalize them.

By standard theory, a set $\{p_k\}_{k=0}^{\infty}$ of monic[12] polynomials, where p_k is of degree k, that are mutually orthogonal on an interval $[a, b]$ with

[12]A monic polynomial is one whose leading coefficient, i.e. the coefficient of the highest power of the variable, is 1.

respect to a weight function $w(x)$ may be constructed by iterating the recurrence relation

$$p_{k+1}(x) = (x - \beta_k)p_k(x) - \gamma_k p_{k-1}(x) \quad \text{for} \quad k \geq 0, \quad p_0(x) = 1,$$

where

$$\beta_k = \langle\, x p_k^2 \,\rangle / \langle\, p_k^2 \,\rangle \quad \text{for} \quad k \geq 0,$$
$$\gamma_k = \langle\, p_k^2 \,\rangle / \langle\, p_{k-1}^2 \,\rangle \quad \text{for} \quad k \geq 1, \quad \gamma_0 = 0.$$

Any desired set of mutually orthogonal polynomials, such as an orthonormal set, can be derived from an appropriate monic set by rescaling.

To turn the above theory into an algorithm it is simply necessary to apply the recurrence relation to calculate p_{k+1} for successive values of k. It is a *two step* recurrence relation, which means that the next value can be computed from the previous *two* values, so that it is never necessary to store more than three successive values. With algorithms like this, it is usually best to use simple scalar variables to represent p_{k+1}, p_k and p_{k-1} for the current value of k, and then to copy p_k to p_{k-1} and p_{k+1} to p_k before incrementing k and computing a new value for p_{k+1}. This avoids potentially expensive accesses to the structure that stores all the computed values of p_k, such as array accesses or, in this case, extracting values from the *end* of a list.

Before executing a loop to compute p_k for general values of k it is necessary to initialize the loop carefully. In any two-step recurrence relation the first two values are special cases that should be handled explicitly before entering the loop. The task of computing the orthogonality integrals is delegated to a support procedure called `DefInt`, which we have seen before (in §4.1). Since all the values used by the integration procedure must be passed as arguments, and cannot (reliably) be taken to be global because of the scoping of variables in algebraic procedures (discussed in §4.7.2), there is no advantage in separating w out of the integrand, and the whole integrand may as well be passed as one argument. As always, care should be taken to try to invent meaningful mnemonic variable names, not to re-evaluate any quantity that has not changed and to use local variables wherever possible.

The following procedure implements the above theory to return, as an algebraic list in order of ascending degree, the set $\{p_k(x)\}_{k=0}^{MaxDegree}$ of monic polynomials that are orthogonal on $[a, b]$ with respect to weight w (assumed to be an expression in x). These two procedures provide a good illustration of algebraic-mode list manipulation.

```
procedure MOP(x, w, a, b, MaxDegree);
   % Make Orthog Polys
   % Construct list of monic polys in x
   % to degree MaxDegree orthogonal wrt. weight w
   % (expression in x) on [a,b].
   begin scalar PolyList, Pkm1, Pk, PkSq, Pkp1,
       IwPkm1Sq, IwPkSq, beta, gamma;
      % Initialize for k = 0:
      % Degree 0 Poly:
      Pk := 1;  PolyList := {Pk};
      if MaxDegree = 0 then return PolyList;
      PkSq := Pk**2;  IwPkSq := DefInt(w*PkSq, x, a, b);
      beta := DefInt(w*x*PkSq, x, a, b) / IwPkSq;
      % Degree 1 Poly:
      Pkp1 := (x - beta)*Pk;  PolyList := {Pk, Pkp1};
      if MaxDegree = 1 then return PolyList;

      return append (PolyList,
         for d := 2 : MaxDegree collect <<
            Pkm1 := Pk;  Pk := Pkp1;  PkSq := Pk**2;
            IwPkm1Sq := IwPkSq;
            IwPkSq := DefInt(w*PkSq, x, a, b);
            beta := DefInt(w*x*PkSq, x, a, b) / IwPkSq;
            gamma := IwPkSq / IwPkm1Sq;
            Pkp1 := (x - beta)*Pk - gamma*Pkm1 >>);
   end$

procedure DefInt(f, x, a, b);
   << f := int(f, x); sub(x=b, f) - sub(x=a, f) >>$

procedure Normalize(PolyList, x, w, a, b);
   % Normalize a list of polynomials defined as for MOP
   % (More efficient to include in MOP as an option.)
   for each Poly in PolyList collect
     ( Poly / sqrt DefInt(w*Poly**2, x, a, b) )$

on failhard;   % No point continuing if integration fails!
```

The switch `FAILHARD` used here causes `INT` to generate an error if it cannot perform an integration, rather than return an unevaluated expression of the form `int(f, x)` as it normally would.

Here is an illustration of the use of the above procedures:

```
% Construct the first few monic Legendre polynomials:
off allfac, ratpri;
Legendre := MOP(x, 1, -1, 1, 3);

                             2            3
legendre := {1,x,(3*x  - 1)/3,(5*x  - 3*x)/5}
```

```
% Normalized they look like this:
Normalize(Legendre, x, 1, -1, 1);

{1/sqrt(2),

 (sqrt(3)*x)/sqrt(2),

               2
 (3*sqrt(5)*x   - sqrt(5))/(2*sqrt(2)),

               3
 (5*sqrt(7)*x   - 3*sqrt(7)*x)/(2*sqrt(2))}

% But Legendre polys proper are scaled so that Pk(1) = 1:
Legendre :=
   for each el in Legendre collect el/sub(x=1, el);

                      2          3
legendre := {1,x,(3*x  - 1)/2,(5*x  - 3*x)/2}
```

Procedure MOP assumes that the REDUCE integrator can evaluate the integrals involved. However, in the fairly common case that the weight function involves square roots this will almost certainly require the ALGINT package, to which we return in chapter 8. For example, the following fails without ALGINT:

```
% Try to construct the first few Chebyshev polynomials:
Cheby := MOP(x, (1-x**2)**(-1/2), -1, 1, 3);

***** failhard switch set
```

4.13 Solving a class of ordinary differential equations

There are many different special types of ordinary differential equation (ODE) for which specialized methods of solution are known. Only certain types have general solutions in "closed form", i.e. expressible in terms of "standard" functions, and in general the solution of an ODE can only be approximated, usually as a truncated power series or numerically. Some of the difficulty of writing a general ODE solver lies in analysing the type of the ODE, and then, to some extent, in decomposing the equation into relevant parts. The ODESOLVE package will be described in chapter 8; here we consider a procedure to solve only the following form of linear ODE with constant coefficients and polynomial "right-hand side":

$$a_n \frac{d^n y}{dx^n} + \cdots + a_1 \frac{dy}{dx} + a_0 y = p(x),$$

where $\{a_i | 0 \leq i \leq n\}$ are constants and $p(x)$ is a polynomial.

Because this ODE is linear in y, its solution may be written as the sum of a complementary function y_0 and a particular integral y_1 in the form

$$y = y_0 + y_1.$$

The complementary function may be constructed as

$$y_0 = \sum_{i=1}^{I} \left(\sum_{j=0}^{J_i - 1} c_{ij} x^j \right) e^{m_i x}$$

where $\{m_i | 1 \leq i \leq I\}$ are the distinct roots of the auxiliary equation

$$a_n m^n + \cdots + a_1 m + a_0 = 0$$

and J_i is the multiplicity of the root m_i.

The particular integral y_1 may be found by assuming that it is a *general* polynomial, substituting $y = y_1$ into the ODE, equating coefficients of powers of x and solving the resulting system of linear equations. If $a_0 \neq 0$, as we will assume for simplicity, then y_1 has the same degree as $p(x)$.

The following procedure attempts to return in the form "`y = ...`" the general solution of `ode`, which is assumed to have *exactly* the form given above. It does not check that `ode` has the correct form, but it does check for the special case of a vanishing right-hand side, which allows it to short-circuit the particular integral code. The argument **`cname`** is the name to be given instead of c to the undetermined coefficients c_{ij} in the complementary function, and is also used similarly for the undetermined coefficients in the particular integral – since the latter need only one argument, representing a subscript, the two uses do not clash.

Let-rules are used to replace derivatives by powers of some unbound identifier in order to construct the auxiliary equation. It is assumed that the auxiliary equation has only real roots, and that the REDUCE operator `SOLVE` is capable of solving in a usable manner all the algebraic equations that arise. In reality, solving the auxiliary equation is probably the biggest difficulty in solving this type of ODE, because in general it would need to be solved numerically for which there is as yet no support built into REDUCE. (However, there is a new user-contributed package for finding all the roots of a complex polynomial, and other work is in progress in this area.)

```
procedure LCCODE(ode, y, x, cname);
   % Solve a Linear Constant Coefficient ODE
   % with a polynomial on the right.
   begin scalar m, auxeqn, y0, y1, roots, rhsode;
      integer i, mlt;
      % Construct auxiliary equation:
      m := symbolic gensym();
         % guaranteed unbound identifier
      for all n let df(y, x, n) = m**n;                 %%
      let df(y, x) = m;                                 %%
      auxeqn := sub(y=1, lhs ode);                      %%
      clear df(y, x); for all n clear df(y, x, n); %%

      % Solve it (ASSUMING Solve can do so!):
      roots := solve (auxeqn, m);

      % Construct the complementary solution y0:
      clear cname;  % to allow tidy re-runs
      operator cname;
      i := 0;
      y0 := for each root in roots sum <<
         root := rhs root;  i := i + 1;
         mlt := first multiplicities!* - 1;
         multiplicities!* := rest multiplicities!*;
         for j := 0 : mlt sum cname(i,j)*x**j >>
         * exp( root*x );

      % Construct a potential particular solution:
      rhsode := rhs ode;
      if rhsode = 0 then y1 := 0 else
         begin integer d; scalar EqnList, VarList;
            d := deg(rhsode, x);
            y1 := for j := 0 : d sum cname(j)*x**j;
            EqnList :=
               coeff(sub(y=y1, lhs ode - rhs ode), x);
            VarList := for j := 0 : d collect cname(j);
            EqnList := solve(EqnList, VarList);
            % Sub accepts (nested) lists of equations:
            y1 := sub(EqnList, y1)
         end;

      return y = y0 + y1
   end$

% Try it, with appropriate display settings:
factor e**x;
on div, intstr;  % on intstr needed in equations

DEPEND Y,X;
ODE := DF(Y,X,2) - 4DF(Y,X) + 4Y = X**2;

                                          2
ode := df(y,x,2) - 4*df(y,x) + 4*y=x
```

```
LCCODE(ODE, Y, X, C);

   2*x                               1  2     1      3
y=e    *(c(1,1)*x + c(1,0)) + ---*x  + ---*x + ---
                                     4        2      8
SUB(WS, ODE);

 2  2
x =x


% Constant rhs:
ODE := DF(Y,X,2) - 4DF(Y,X) + 4Y = A;

ode := df(y,x,2) - 4*df(y,x) + 4*y=a

LCCODE(ODE, Y, X, C);

   2*x                               1
y=e    *(c(1,1)*x + c(1,0)) + ---*a
                                     4
SUB(WS, ODE);

a=a

off div, intstr;  remfac e**x;
```

It is possible that the operation of the four lines marked on the right with `%%` in `procedure LCCODE` above is implementation dependent. If so, it may be necessary to swap the order of the two let-rules; if that does not work then a modification that should always work would be to use each rule and then clear it immediately, as described in §2.8 and illustrated in §2.9. However, an altogether more elegant way to construct the auxiliary equation is to replace the four lines marked with `%%` by simply

```
auxeqn := sub(y=exp(m*x), lhs ode/y);
```

but this does not illustrate the use of let-rules within procedures!

The REDUCE code follows the standard pencil-and-paper algorithm very closely (as any first implementation of an algorithm should aim to do). The unknown m in the auxiliary equation has to be an unbound identifier, so we have assigned to the local variable `m` an unbound identifier generated by calling the symbolic-mode function `GENSYM`. This reliably avoids the name clash that might arise if we just used `m` (or any other explicit identifier) as a global variable directly. The only reason for clearing `cname` before declaring it as an operator is to avoid a warning message when the procedure is re-run. The following points of interest arise in constructing the particular solution. COEFF can only be applied to expressions, not equations, so ode must be turned explicitly into the expression "LHS ode - RHS ode". This is actually done inside SUB, although SUB can be used to substitute into an equation just as well as into an expression, and we have used the fact that not only can variables that evaluate to equations be used as all but the last argument to SUB, but so can possibly nested lists of equations. The list of variables to be used by SOLVE is constructed using a FOR ... COLLECT loop.

We conclude this example with some technical remarks. As we warned earlier, if procedure LCCODE is compiled it will probably be necessary to precede the procedure definition by the symbolic-mode declaration

```
FLUID '(X Y M);
```

in order that the X, Y, M appearing in the let-rules are recognized as respectively procedure arguments and a local unbound identifier. For tidiness, the procedure definition can be followed by

```
UNFLUID '(X Y M);
```

In fact, there is probably little to be gained by compiling this procedure, but some REDUCE implementations will be set up to compile by default. This procedure may fail for an ODE with a symbolic constant alone on the right, due to a bug in the REDUCE function DEG – see the Appendix, §D, for more details. Also, note that for ON DIV (and most other output switches) to be obeyed within an equation it is necessary to set ON INTSTR.

4.14 Solving coupled nonlinear equations numerically

It may seem odd to use a computer algebra system to solve a problem numerically, but most systems have very powerful numerical capabilities – more powerful in terms of precision than conventional languages like FORTRAN, but consequently much slower. Nevertheless, for once-only solutions of numerical problems that have a significant algebraic content it may be very convenient to use an algebra system. One such problem is implementing the Newton-Raphson algorithm for improving an approximate solution of a nonlinear equation or system, which involves finding and evaluating derivatives. When this is programmed in a language like FORTRAN it is usually left to the user to supply a subprogram that evaluates both the function itself and its derivative at a specified point, so that the user is left with the task of differentiating the function before the algorithm can be programmed. If the function is complicated one may well seek the assistance of an algebra system to make sure that the derivative is correct, and perhaps also to generate the FORTRAN code automatically so as to avoid any transcription errors (we return to this possibility in chapters 8 and 9).

The alternative is to have the algebra system solve the whole problem. The case of a single nonlinear equation in one unknown is trivial, so let us consider the n-dimensional problem, which is a straightforward generalization. The theory is as follows. Let $f : \mathbb{R}^n \to \mathbb{R}^n$ be a nonlinear mapping and $\alpha \in \mathbb{R}^n$ be a simple zero of f, i.e. a simple root of the system of n nonlinear equations in n variables

$$f(x) = 0, \quad x \in \mathbb{R}^n.$$

Let $J(x)$ be the Jacobian matrix of f at x, i.e.

$$J_{ij}(x) = \frac{\partial f_i}{\partial x_j}(x),$$

and let $J^{-1}(x)$ be its inverse, which can be shown to exist for x near to a simple zero of f.

If $x_{(0)} \in \mathbb{R}^n$ is sufficiently close to α then it can be shown that the sequence $\{x_{(r)}\}$ defined by

$$x_{(r+1)} = x_{(r)} - \text{shift}(x_{(r)}), \quad \text{where} \quad \text{shift}(x) = J^{-1}(x)f(x),$$

converges to α. This is Newton's algorithm in n dimensions. The absolute error $e_{(r+1)}$ in the approximation $x_{(r+1)}$ may be estimated by using

$$e_{(r+1)} \approx \|x_{(r+1)} - x_{(r)}\| = \|\text{shift}(x_{(r)})\|,$$

where $\|\cdot\|$ is a suitable vector norm such as the Euclidean norm – the geometrical length of the vector – defined by

$$\|x\| = \left(\sum_{i=1}^{n} x_i^2\right)^{1/2}$$

The following procedure implements this theory to find the root of $f(x) = 0$ to which Newton's algorithm converges from the initial approximation x_0, accurate to within an absolute error `AbsErr` estimated using the Euclidean norm. The arguments `f`, `x` and `x0` are n-element algebraic lists used to represent n-vectors. As a sensible precaution, the procedure checks that they all have the same dimension (i.e. length) n.

The list `x` contains the names of the variables used in the expressions in `f`.

The root is returned as an n-element algebraic list, with the elements ordered as in `x` and `x0` (by analogy with the built-in operator `SOLVE`). Matrix arithmetic is used internally within the procedure, which must convert from the above list representation to matrix representation as necessary. Bigfloat number representation is used internally, and `AbsErr` and the elements of `x0` are assumed to be in bigfloat representation. The procedure sets the bigfloat precision to a suitable value using a technique that is not very elegant but uses only algebraic mode. This, and other features of this procedure such as converting between list and matrix representation, could probably be done much more neatly with a little symbolic-mode code (to be introduced in subsequent chapters).

This implementation computes the value of the variable `shift` once only as a matrix with symbolic elements, and then evaluates it numerically at the latest estimate of the root on each iteration until its norm is small enough. This is done by recursively substituting into each element numerical values for all the variables in the "vector" `x`.

```
procedure NewtonN(f, x, x0, AbsErr);
   % f, x & x0 are n-element lists representing vectors
   % Solves f(x)=0, starting at x=x0, to absolute accuracy
   % AbsErr, using multi-dimensional Newton iteration,
   % returning the solution in a list matching x & x0.
   begin integer n, i, j, Prec;  real absvar0, RootMax;
      % local numval switch and bigfloat precision:
      scalar !*numval, OldPrec;
      OldPrec := precision 0;  % current precision
      if (n := length f) neq length x or n neq length x0
         then rederr
           "NewtonN: Lists do not have the same lengths";

      % Set up the pseudo-local matrices:
      matrix NewtonN!-ff(n,1), NewtonN!-Ja(n,n),
         NewtonN!-Root(n,1), NewtonN!-shift,
         NewtonN!-numshift;
      i := 0;  for each fun in f do <<
         i := i + 1;
         % Assign expressions for f to matrix ff:
         NewtonN!-ff(i,1) := fun;
         % Assign the Jacobian to matrix Ja:
         j := 0;  foreach var in x do
           NewtonN!-Ja(i,j:=j+1) := df(fun, var)
         >>;
      % Set up the symbolic shift matrix:
      NewtonN!-shift := NewtonN!-Ja**(-1) * NewtonN!-ff;
      clear NewtonN!-Ja, NewtonN!-ff;
```

```
on numval, bigfloat;
% Assign initial approximation to matrix Root and
% estimate magnitude of root (using infinity-norm):
Rootmax := 0;
j := 0;  for each var0 in x0 do <<
   NewtonN!-Root(j:=j+1,1) := var0;
   absvar0 := abs var0;
   if absvar0 > RootMax then RootMax := absvar0
   >>;

% Estimate required precision as
% Prec := Ceiling (log10 RootMax - log10 AbsErr)
% where log10 a = log a / log 10:
RootMax := if abs RootMax <= e then 1
               else log RootMax;
RootMax := (RootMax - log AbsErr) / log 10.0;
Prec := 1; while Prec < RootMax do Prec := Prec + 1;
precision Prec;

AbsErr := AbsErr**2;  % minor efficiency hack!
repeat <<
   NewtonN!-numshift := NewtonN!-shift;
   for i := 1:n do <<
      j := 0;  for each var in x do
         NewtonN!-numshift(i,1) :=
            sub(var=NewtonN!-Root(j:=j+1,1),
               NewtonN!-numshift(i,1));
      >>;
   NewtonN!-Root:= NewtonN!-Root - NewtonN!-numshift
   >> until
      (for i:=1:n sum NewtonN!-numshift(i,1)**2)
         < AbsErr;

   x0 := for i := 1:n collect NewtonN!-Root(i,1);
   clear NewtonN!-Root, NewtonN!-shift,
      NewtonN!-numshift;
   precision OldPrec;
   return x0
end;
```

The matrix variables used in this procedure have been given obscure names to avoid potential conflicts, since they are global. A neater solution would have been to treat them properly as local objects using the techniques that will be introduced in chapter 9, which again involve symbolic-mode code. The conversion between list and matrix representations uses the techniques that were discussed in chapter 3. An alternative approach would have been to use the PART operator to access the lists via an index.

The heart of the algorithm is the three statements in the REPEAT loop (to which we will refer without the prefix "NewtonN!-"). The statement

```
Root := Root - numshift
```

implements the Newton-Raphson iteration and makes use of matrix algebra to do this succinctly. The two statements before this update the value of `numshift` in two steps:

```
numshift := shift
```

makes `numshift` purely symbolic, and then the double for-loop replaces each variable in the variable list `x` in each element of the column matrix `numshift` by its latest numerical value stored in the column matrix `Root`.

Estimating the correct precision to use means estimating the number of significant decimal digits required in the final answer. This would have been slightly easier in terms of relative rather than absolute error, but relative error has the disadvantage that it may not be appropriate for values near zero, and of course fails completely at zero. In order to use more than enough significant figures, the procedure uses the ∞-norm (which takes the magnitude of the largest component) to estimate the magnitude of the root, which it assumes to be the same as the magnitude of the initial approximation. This gives a value to the variable `RootMax`, which is computed in the same loop that assigns the initial values from the list `x0` to the matrix `Root`. Then the precision is estimated as the ceiling of (i.e. the smallest integer not less than) $\log_{10}$(`RootMax/AbsErr`).

The ceiling function is implemented by simply counting up from 1, which should not be grossly time-consuming for a precision likely to be of order 10, and requires only algebraic mode. A heuristic is used to avoid problems caused by small values of `RootMax`, particularly zero, since $x_{(0)} = 0$ is a fairly likely initial approximation. (The procedure assumes without checking that the user will not be so unwise as to specify a maximum error of zero!)

Having explained how it works, let us test the procedure on a very simple problem:

```
F := {X - COS Y / 2, Y - SIN X / 2};

         cos(y) - 2*x       sin(x) - 2*y
f := { - -------------- , - --------------}
               2                  2

ON BIGFLOAT;
ROOT := NEWTONN(F, {X, Y}, {0, 0}, 0.001);

root := {0.486 406,0.233 726}
```

```
% Takes about 25s + similar GC time on 1M Atari ST at 85%
% store usage.  Compiling makes no significant difference!

% Check it:
X := FIRST ROOT$  Y := SECOND ROOT$
ON NUMVAL;  F;

{0.000 00090 3008,0.000 00012 43955 }
```

The value of f is 0 to within the requested accuracy. An interesting observation was that compiling this procedure did not produce any significant change in speed on this problem. The reason is probably that most of the work is being done by calls to REDUCE procedures that are themselves already compiled, and the overhead of interpreting the code that "organizes" these procedures is negligible. It is symbolic procedures particularly that normally benefit from compilation.

4.15 Local switch and domain setting

We conclude this chapter with a brief introduction to the techniques that are necessary to preserve the global environment in which a procedure (or block) is executed, the simplest of which were used in `procedure NewtonN` above.

The identifier of the internal logical variable corresponding to a switch called `SWITCH` is `!*SWITCH`, and declaring such an identifier to be a local variable means that its local value is used in preference to its global value, leaving the latter unchanged. This technique works for the logical variable representing the switch itself, but that may not be sufficient, because changing switch settings using `ON` and `OFF` can have side effects that are not handled by this technique; it would require a small amount of symbolic-mode code to handle "local switch settings" properly, as illustrated in the example below.

The side effect of setting some switches is just to clear the simplified indicator on all the algebraic values used so far (see chapter 5 for further details). This is the case for the switch `NUMVAL`. It may well actually be desirable to effectively reset the `NUMVAL` switch when leaving a procedure *without* clearing the simplified indicator if it is appropriate to preserve any value returned by the procedure without converting it according to the current settings in the global environment. This will often be the case with `NUMVAL`, as in `procedure NewtonN` above, and then the local switch variable technique can safely be used.

Unfortunately, the analogous trick for the `BIGFLOAT` switch essentially fails, because the principal effect of setting domain-mode switches is to change the global domain mode, and the logical switch variable itself is hardly used at all. We illustrate how to handle local domain setting below. In `procedure NewtonN`, `BIGFLOAT` was simply turned on globally and left on, partly to avoid undue complication, and partly because this is almost certainly the correct mode to have in effect when the procedure returns, even if it was not the mode when the procedure was entered.

The local variable technique also does not work for the bigfloat precision, because in this case the relevant variable (`!:PREC!:`) has the wrong internal type (`global` rather than `fluid`). However, `PRECISION 0` returns the current precision without changing it (see §1.10), so this can be used to preserve the precision explicitly and later restore it, using entirely algebraic-mode code as in `procedure NewtonN` above.

The following procedure illustrates a technique for using bigfloat mode locally within a procedure without changing the global domain mode; as mentioned above it requires a small amount of symbolic-mode code (see chapters 5, 6 and 9 for further details). The switch `NUMVAL` is handled in exactly the same way that it was in `procedure NewtonN` as described above. Bigfloat mode is established by setting the domain mode directly, and not by using the `BIGFLOAT` switch at all. The internal function `setdmode` returns the *tag* of the current domain mode, or `nil` if it is the default mode; the actual domain name can be obtained from the tag as the value of its `dname` property.

```
procedure bf val;
   % Returns val evaluated in bigfloat mode
   % WITHOUT CHANGING THE GLOBAL ENVIRONMENT.
   begin scalar mode, !*numval;
      on numval;  % also clears simplified indicator
      % Set mode to bigfloat, and save previous mode:
      off msg;  % to avoid a possible warning message
      symbolic(mode := setdmode('bigfloat, t));  on msg;

      % Do the required algebraic computations -
      % in this simple example just:
      val := val;

      % Turn bigfloat off locally, and
      % restore previous mode if not default:
      symbolic << setdmode('bigfloat, nil);
         if mode then setdmode(get(mode, 'dname), t) >>;
      % Simplified indicator not cleared,
      % so val will stay in bigfloat mode:
      return val
   end$


% Check that it works, and that the returned value does
% not get re-simplified into the current domain mode:
```

```
BFPI := BF PI;

bfpi := 3.141 59265 4

BFPI;

3.141 59265 4

% This is because the simplified indicator is set -
% see chapters 5 and 6:
GET('BFPI, 'AVALUE);

(scalar (!*sq ((!:bf!: 314159265359 . -11) . 1) t))

% Check the environment is still the default:
PI;

pi

0.5*1;

*** 0.5 represented by 1/2

 1
---
 2
```

The model used above for BIGFLOAT should work for all domain modes, although it is possible that in some situations it may be necessary to set the appropriate switch variable as well. Most other switches should be reliably handled by the model used for NUMVAL. If necessary, the simplified indicator can be explicitly cleared by

```
symbolic rmsubs();
```

4.16 Exercises

1. Write procedures called MAX and MIN to take a single argument that should be an algebraic list of numerical values, and return respectively their maximum and minimum values. The procedures should work in all number domains (if they are written as algebraic procedures in the obvious way then they should automatically do so). Add code to handle invalid arguments in some appropriate way. Can you arrange that if any arguments are symbolic then they remain symbolic, and only numerical arguments are actually extremized, using only algebraic-mode code? [Probably not!] We return to this problem in chapter 9 and its exercises, where we indicate how to solve this problem completely

– it will be useful to save for then the procedures you write now.

Just for fun, try writing these procedures both iteratively (which is probably best) and also recursively. See if you can find an efficient way to make most of the code common between `MAX` and `MIN`, since all but the order relation will be identical. Consider code-sharing again when you reconsider this problem after chapter 9.

2. Write and test a recursive procedure to compute large integer powers (primarily of numerical values) using the algorithm

$$x^n = \text{if } n \text{ even then } (x^{n/2})^2 \text{ else } x(x^{(n-1)/2})^2,$$

to which you must add a suitable stopping condition or base case to avoid infinite recursion.

(You might also like to work out the computational complexity of this algorithm, i.e. the number of actual multiplications, and compare it with the $n-1$ multiplications required by the straightforward approach.) But is this algorithm necessarily better for very large numbers and expressions, both of which require list representations?

3. The binomial expansion may be defined as

$$(1+x)^n = \sum_{r=0}^{\infty} {}^nC_r\, x^r$$

for any x and n, where

$${}^nC_r = \frac{n(n-1)(n-2)\cdots(n-r+1)}{1\cdot 2\cdot 3\cdots r}.$$

If n is a non-negative integer the expansion terminates after the term with $r = n$, because clearly

$${}^nC_r = 0 \quad \text{for} \quad n < r.$$

Write a body for

```
procedure BinExp(x, n, MaxDeg)
```

which returns the binomial expansion of $(1+x)^n$ up to maximum degree `MaxDeg` *efficiently*. It should not compute any term that must vanish. If `MaxDeg` is not a non-negative integer the procedure should report an error and return *unless* n is a non-negative integer, in which case `MaxDeg` should be ignored.

Use this procedure together with other REDUCE facilities to assign to the variable `z` the power-series expansion of:

(a) $(1+x^2)/(1-x^2)$ about $x=0$ up to degree 10;

(b) $(1+x+y^2)^{-2}$ up to total degree 10 in x and y, i.e. including only power products of the form $x^\alpha y^\beta$ such that $\alpha+\beta \leq 10$.

If the expansions of $\sqrt[3]{1+px}$ and $(1+2qx)/(1+qx)$ about $x=0$ agree up to second degree in x show that $p=3q$, where p and q are constants.

4. Write more complete versions of the `DOT` and `CROSS` operators introduced in the text, in particular so that dot and cross products of vectors expressed as linear combinations of orthonormal basis vectors are simplified correctly. Try to provide support for triple products, particularly to incorporate the symmetry of the scalar triple product. [This may not be easy.] Add some support for vector calculus: grad, div, curl, etc.

5. Define non-commuting position (q) and momentum (p) operators with the following commutation rules in 3 dimensions (so that $1 \leq r, s \leq 3$) and check them for a few explicit values of r and s:

$$\begin{aligned} q(r)q(s) - q(s)q(r) &= 0 \\ p(r)p(s) - p(s)p(r) &= 0 \\ q(r)p(s) - p(s)q(r) &= i\hbar\delta_{r,s} \end{aligned}$$

where

$$\delta_{r,s} = \begin{cases} 1 & \text{if } r = s \\ 0 & \text{otherwise} \end{cases}$$

as introduced in chapter 3, exercise 9.

6. Experiment with using let-rules to differentiate and integrate polynomials term by term, i.e. define *linear* operators to differentiate and integrate x^n for all $n \geq 0$. You will need to consider all the cases that REDUCE might consider as special, even if they are not special mathematically.

7. If $y = (1 - x^2)^{-1/2} \arcsin x$, prove that

$$(1 - x^2)y^{(1)} = xy + 1.$$

By differentiating this equation n times using Leibniz' formula, show that

$$(1 - x^2)y^{(n+1)} - (2n + 1)xy^{(n)} - n^2 y^{(n-1)} = 0.$$

When $x = 0$, obtain $y^{(0)} = 0$ and $y^{(1)} = 1$. Show that, in general, $y^{(n+1)} = 0$ (n odd) and $y^{(n+1)} = 2n\{(\frac{n}{2})!\}^2$ (n even).

8. Add support for the nth derivative of x^n to the DFN code, and then by differentiating $x\,dy/dx = x^n + ny$ symbolically n times using Leibniz' formula, *deduce* that $d^{n+1}y/dx^{n+1} = n!/x$.

9. Generalize **procedure** LCCODE to remove the restriction $a_0 \neq 0$, allow complex roots of the auxiliary equation, allow more general right-hand sides with exponential and trigonometric functions, etc.

10. Picard's method for generating an approximate algebraic solution of the ODE

$$\frac{dy}{dx} = f(x, y), \quad y(x_0) = y_0,$$

is to iterate the formula

$$y_{n+1}(x) = y_0 + \int_{x_0}^{x} f(x', y_n(x'))\,dx'$$

starting from $y_0(x) = y_0$. Write a body for

```
procedure Picard(f, x, y, x0, y0, d)
```

where f is a polynomial in x and y, to return the approximate solution $p(x)$ of the ODE that agrees with the true solution $y(x)$ *up to degree* d, i.e. the polynomial $p(x)$ that satisfies

$$y(x) - p(x) = O(x^{d+1}).$$

The whole computation should be performed only up to degree d, in which case the stopping condition is when two successive approximations are equal. Note that there is no guarantee that all the terms of an approximation $y_n(x)$ will be correct – typically one or more of the higher degree terms will have incorrect coefficients.

Check your procedure first by taking $f(x, y)$ to be a low degree polynomial in x alone, and then to be ay. Then use it to solve

$$\frac{dy}{dx} = 1 + 3xy^2, \quad y(0) = 1$$

to degree 6 and check the result *carefully*.

11. This question is about multivariate Taylor expansion. Vectors in n-dimensions are best represented in REDUCE 3.3 as lists (whereas in previous versions the best representation was probably as arrays, with the dimension assigned to the 0th element for easy access). Write a body for

```
procedure TaylorN(f, x, x0, d)
```

to return the Taylor expansion of f with respect to x about x_0 up to total degree d, where x and x_0 are n-dimensional vectors and f is an expression depending on the elements x_i of x. That is, the expansion should include all monomials

$$\prod_{i=1}^{n} x_i^{d_i} \quad \text{with degree} \quad \sum_{i=1}^{n} d_i \leq d.$$

In order to explain possible algorithms we introduce the following notation, which does not necessarily need to be used to write the procedure. Let

$$T_n(f, x, x_0, d)$$

denote the *multivariate* Taylor expansion of f with respect to the first n elements of x about x_0, and let

$$T[f, z, z_0, d]$$

denote the *univariate* Taylor expansion of f with respect to z about z_0, in both cases up to degree d.

There are two obvious ways to compute a Taylor expansion in n variables:

1. Use the recursive definition

$$T_n(f, x, x_0, d) = \{T[T_{n-1}(f, x, x_0, d), x_n, x_{0n}, d]\}_d$$

$$T_0(f, x, x_0, d) = \{f(x)\}_d$$

where the notation $\{expression\}_d$ means truncate the polynomial expression above total degree d. (When implementing this algorithm using list representation it may be more convenient to run through the variables in the opposite order to that implied above.)

2. Use the definition

$$T_n(f, x, x_0, d) = T[f(\ell(x - x_0) + x_0), \ell, 0, d]\Big|_{\ell=1}$$

i.e. define the univariate function $g(\ell) \equiv f(\ell(x - x_0) + x_0)$, expand it with respect to ℓ about 0 up to degree d, and then set $\ell = 1$.

In both cases the problem is reduced to that of univariate Taylor expansion. Implement both definitions and compare their relative ease of programming and running speed. As examples, compute the expansions of

$$f_1(x, y) = \sin(x) + \sin(y) \quad \text{and} \quad f_2(x, y) = \sin(xy)$$

both about $x = 0$ as univariate functions and about $x = y = 0$ as bivariate functions, up to total degree 10.

12. Write a procedure to construct the composition of two maps

$$\mathbb{R}^\ell \to \mathbb{R}^m \quad \text{and} \quad \mathbb{R}^m \to \mathbb{R}^n.$$

As a safety precaution, the two functions should be allowed to use the same variable names, so that for example $(f_1(x, y), f_2(x, y))$ could be composed with $(g_1(x, y), g_2(x, y))$.

13. (Suggested by Charles Leedham-Green.) Let v_0, v_1, v_2 be the vertices of a triangle inscribed in a circle of unit radius centred at the origin. Let p be any point on the circumference, and let q_0, q_1, q_2 be the feet of the perpendiculars from p to the three sides of the triangle.

(a) Prove that q_0, q_1, q_2 are collinear.

(b) Find the envelope of the line through q_0, q_1, q_2 as p varies (difficult, and really needs graphics).

14. L'Hopital's rule states (roughly) that if $f(a) = g(a) = 0$ or ∞ and $f(x)$ and $g(x)$ are both differentiable near $x = a$, then

$$\lim_{x \to a} \frac{f(x)}{g(x)} = \lim_{x \to a} \frac{f'(x)}{g'(x)}$$

and so on (see e.g. Abramowitz and Stegun 1964, §3.4.1). Write a body for

```
procedure limit(fun, x, a)
```

to take a limit using l'Hopital's rule. You may implement this either recursively or iteratively, but you will find that an iterative implementation runs faster. You should set a limit of about 20 on the depth of recursion or number of iterations. `Limit` should return the identifier `infinity` if it encounters "non-zero/zero". Test your routine with limits such as $(\sin x)/x$ as $x \to 0$. Try $\sqrt{(\sin x)/x}$ as $x \to 0$ – why does this not work? How might one write a procedure that could evaluate such a limit? Think about this again when you know more about symbolic mode.

A function such as $(\sin x)/x$ has a perfectly well defined Taylor expansion about $x = 0$. Modify the Taylor routine given in the text so that it correctly evaluates such expansions.

15. The *leading order* of a function $f(x)$ at x_0 may be defined as the value of p such that the limit as $x \to x_0$ of $f(x)/x^p$ is a non-zero constant. Thus it captures the power-law behaviour of $f(x)$ in the neighbourhood of x_0. Convince yourself that it is given by

$$\lim_{x \to x_0} \frac{(x - x_0)f'}{f}.$$

Write and test a body for

```
procedure leading!-order(f, x, x0)
```

to return the leading order of f with respect to x at x_0.

16. If the function $f(x)$ has a pole of order p at the point x_0 then its leading order is a negative integer and its leading term has the form $c(x - x_0)^{-p}$ for some constant c so that $(x - x_0)^p f(x)$ is regular at x_0, and the Laurent series expansion of $f(x)$ is

$$(x - x_0)^{-p} T[(x - x_0)^p f(x)]$$

where T denotes the Taylor series operator. By using the routines developed in the previous question and in §4.9, write

```
procedure Laurent(f, x, x0, n)
```

to return the Laurent expansion of $f(x)$ about x_0 up to degree n (take care about this degree) in x, and assign the order p of the pole to the global variable `POLE!-ORDER`. Apply this procedure to find the expansion of

$$f(x) = \cos(x)/x^3$$

about the origin up to degree 5.

In principle, the Laurent routine developed above should cope with expanding about 0 a function such as

$$f(x) = \cos x \operatorname{cosec} x^3.$$

Experiment with this, and try to develop a routine that handles this kind of expansion effectively.

17. Write a package for handling extendable series, e.g. a Taylor series procedure that can be called to generate some more of an expansion without re-computing what is already known. [This could be thought of as an extension of the factorial problem in chapter 3, exercise 10. The recent TPS package described in chapter 8 actually provides this facility.]

5

A look inside REDUCE

5.1 Introduction: some abstract algebra

To understand what we find when we look inside an algebra system we must first have some idea what the data is that the system is representing. For emphasis, we repeat here that although the systems are performing calculus, they are *not* doing analysis; that is, they are emphatically *not* taking limits, testing convergence, and so on, except where those operations can be done by a strictly algebraic method: this is of course just what humans do when performing calculus. As mentioned in the first chapter, all of us will happily differentiate elementary functions without going back to first principles and taking limits, because we know the computational rules, like the rule for differentiating a product, which ensure that the results are correct, and we know that all the usual functions obey these rules. The rules are all in fact algebraic in nature.

At this point (and for chapter 7) it helps to have somewhat more formal definitions of the algebraic structures than we have needed up to now. We will give these in an intuitive fashion rather than in full detail, but readers may still find them a little heavy compared with the more practical parts of this book. A useful and more detailed introduction to the algebra needed for algebraic computing is provided in the book by Lipson (1981).

Definition: A *ring* is a set R together with the operations $+$, $-$ and $*$ obeying the usual rules of addition and subtraction (e.g. it contains an element, usually written 0, such that $a + 0 = a$ and $a * 0 = 0$ for all elements a in R) and obeying the closure, associative and distributive laws for multiplication. However, the multiplication need not be commutative, e.g. it could be multiplication of matrices, and it need not have a multiplicative identity, i.e. an element, usually denoted 1 if it exists, such that $1 * a = a$ for all a in R. We shall assume the rings we talk about do have such an element.

A general ring need not obey all the rules of arithmetic because division need not be possible. The ring $\mathbb{Z}$ of integers gives an example, since the division of one integer by another does not always yield an integer. Rings with commutative multiplication which allow division are called fields.

Definition: A *field* is a set of elements together with definitions of all the normal processes of arithmetic (i.e. $+$, $-$, $*$ and $/$) such that the set obeys the usual rules we expect for those processes.

An example of a field is given by the set $\mathbb{Q}$ of rational numbers, with the usual arithmetic operations. One can prove that a ring with commutative multiplication in which there are no non-zero elements a and b such that $a * b = 0$ (such rings are called *integral domains*) has an associated *field of fractions* (or *quotient field*) related to the ring just as rational numbers are related to integers. In practice, users of REDUCE are mainly interested in rings and fields which are constructed by starting from one of the following: the ring of integers $\mathbb{Z}$, and the fields given by the rational numbers $\mathbb{Q}$, the real numbers $\mathbb{R}$, the complex numbers $\mathbb{C}$ and the finite fields $GF(p)$ for each prime integer p. (The definition of $GF(p)$ will be given in chapter 7, when we need it.)

An important construction in the theory of rings is that of polynomials. When defining polynomials we can avoid the somewhat mysterious introduction of an "unknown" by working in terms of sequences of elements of a ring R (which will be the coefficients in the polynomial) of which only a finite number are non-zero. Addition and multiplication of sequences can then be defined so that they will agree with the familiar algebra of polynomials if we use as the "unknown" the element represented by the sequence $\{a_i\}$ in which only $a_1 = 1$ differs from zero. Calling this element x we then have a new ring $R[x]$ of polynomials in x with coefficients in R. By definition, all polynomials have finite degree. (The notation $R[[x]]$ or $R\langle x\rangle$ is used for the ring of all power series in x, including infinite ones.) Now, if $R[x]$ is an integral domain, we can define the field $R(x)$ of rational functions of x (i.e. functions whose numerator and denominator are polynomials). This can always be done if R is an integral domain; if R is a field F, $F(x)$ is called the extension of F by x.

We can pile these constructions on one another. For instance we can define the polynomial ring $S[y]$ and quotient field $S(y)$ for the case where S is itself a polynomial ring $R[x]$. This gives polynomials in more than one variable (multinomials), and the rational functions built from them: the resulting ring and, when relevant, quotient field are denoted by $R[x, y]$ and $R(x, y)$. Thus, starting from integers (or real or complex numbers if necessary), and writing each function as a rational whose numerator and denominator are each polynomials in one or more variables (including here the transcendental functions, e.g. regarding $\sin x$ as an additional variable extending $R(x)$, rather than as a function of x), we can define all the usual functions in terms of rational functions (i.e. elements of $R(x)$) and a few basic functions such as exponentials and logarithms.

The most fundamental capabilities of algebra systems thus rest on the handling of polynomials, which is why texts on the subject (e.g. Davenport *et al.* 1988) dwell on the resulting problems at length. It is helpful, but not essential, for efficiency if we have a fixed form for a given polynomial. The significance is that then the comparison of polynomials is easy: one has only to compare the patterns of bits held in the computer without doing any manipulation which entails an understanding of the meaning of those patterns as algebraic values. Such representations are called *canonical*. For polynomials in one variable, an obvious canonical representation is one in which all terms of the same degree are collected into one, and the terms of different degrees are ordered by descending (or ascending) degree. For multinomials there are similarly obvious representations: one can for instance think of a multinomial as a polynomial in a first (main) variable with coefficients which are polynomials in the remaining variables, and then treat these coefficients recursively in the same manner. REDUCE's default is such a recursive representation, as we will describe later (cf. remarks in §2.6), but REDUCE allows the user to set switches which affect the precise form.

Before giving details of the representation used by REDUCE, we briefly describe the available alternatives. Instead of writing a polynomial as a series of terms in decreasing (or increasing) degree, one could first factorize it and then order the factors in some way: so one has a choice of factored or expanded forms. Secondly, with a multinomial one may prefer to order the products of the variables so that all terms of a given total degree appear together: for instance one may prefer

$$1 + x + 2y + x^2 + 4xy + 5y^2$$

to the recursive form

$$(1 + 5y^2) + (1 + 4y)x + x^2.$$

The first of these is called a distributed form.

Having decided how the polynomials would be written on paper, one has to decide how to represent them in the computer. Here the choice rather depends on the usage: if one is always dealing with polynomials almost all of whose terms are present it is worthwhile to list all possible powers and their coefficients (this is called a dense representation). With multinomials this is usually not the case and it becomes more efficient to use a sparse representation in which only those terms actually present are recorded. For example, writing $1 + x^5y^7$ in a way that recorded explicitly that each of the coefficients of terms involving powers of x lower than 5 or of y lower than 7 was zero would be very wasteful. It is the solutions adopted to questions like these which decide what tasks a specific algebra system is best adapted to, and make it almost impossible to declare any one of them "the best". [It is worth noting that the efficiency of recursive computations may be heavily affected by the ordering of the variables or of the indexing, just as index ordering affects efficient handling of large arrays in many programming languages (Pearce and Hicks 1981, Gardin and Campbell 1983).]

5.2 Lisp data structures

5.2.1 Lisp and REDUCE

The reason for including Lisp in this book is that it is the language underlying REDUCE, as stated in §1.2. It also underlies other computer algebra systems, especially MACSYMA and muMATH (see chapter 1). In fact several systems share a common ancestor in Martin's MATHLAB from MIT in the early 1960s.

By looking at this lower level of REDUCE we can understand how REDUCE actually carries out its manipulations: one notable aspect of algebraic computing is that although the main user language is very high level, advanced use requires one to understand rather more of what goes on internally than is the case for most other kinds of computation. (Some of the remarks here repeat and expand on remarks in earlier chapters.)

A major difficulty in teaching anything general about Lisp is that it is very much *not* a standardized language: until recently almost every machine which had Lisp had a different Lisp. Figure 5.1 shows a Lisp family tree – very much simplified.

Cambridge Lisp has been described by one of its designers (John Fitch) as derived from Lisp 1.8 + 0.3i (to go round the singularity of Lisp 2.0 which looked like ALGOL) and about 4 other Lisps. It began life as an IBM mainframe program, and that still shows in places!

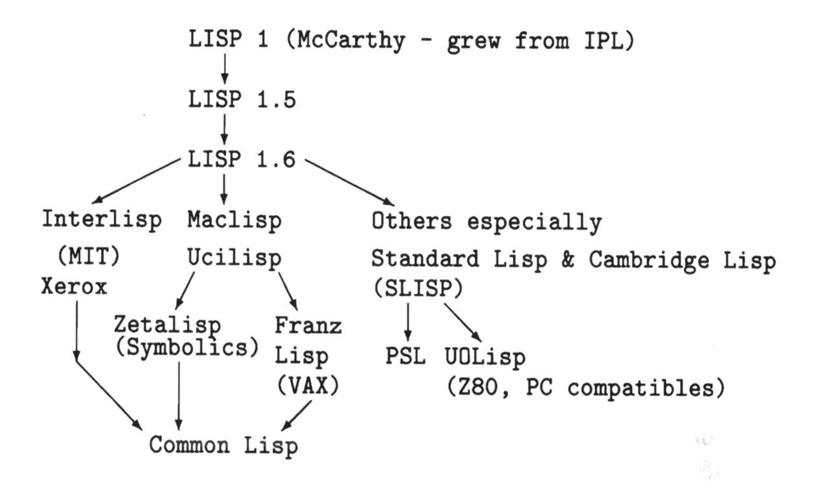

Fig. 5.1. A Lisp family tree

MACSYMA originally used the Maclisp family, and later Common Lisp. These are much bigger Lisps than the Standard Lisp (SLISP) family; they were designed for work on artificial intelligence. SLISP (Marti *et al.* 1979) was intended to be small and efficient, especially as a basis for algebra and other systems. An oversimplified description would be to say that Common Lisp is the union of all the features anybody wanted in a Lisp system, while SLISP is the intersection.

REDUCE uses RLISP, a member of the SLISP family with a syntax based on that of ALGOL-60 (in some ways like Pascal). The symbolic mode of REDUCE consists of RLISP supplemented by additional symbolic mode commands defined in the parts of REDUCE which provide the algebraic capabilities, so that in symbolic mode the user's input is interpreted like an RLISP command.

To enter symbolic mode, one simply gives the REDUCE command

```
SYMBOLIC;
```

and to return to algebraic mode the command is

```
ALGEBRAIC;
```

(In REDUCE 3.3, `LISP` is a synonym for `SYMBOLIC`, but Anthony Hearn prefers `SYMBOLIC` and we have therefore used this except for cases where we refer to features specific to particular native Lisps, where we emphasize that by using `LISP`. Symbolic mode provides uniform access to any functions defined in the native Lisp that are additional to those defined in SLISP or RLISP, unless they were explicitly removed or redefined when REDUCE was built.)

SYMBOLIC and ALGEBRAIC can also be used as if they were operators (which they are not, in the sense that neither is declared as an operator in either mode) in order to force an individual command or component of a command to be interpreted in a particular mode. For example, in algebraic mode,

```
SYMBOLIC x := y;
```

is interpreted as an algebraic assignment of the algebraic value of y as the algebraic value of the symbolic (Lisp) value of x (because as usual the prefix "operator" SYMBOLIC takes precedence over the infix operator ":="), whereas

```
SYMBOLIC (x := y);
```

assigns the symbolic value of y as the symbolic value of x. As a convenience, any function whose first argument is preceded by the quote mark ' is processed in symbolic mode (see §6.1). The name of the current mode is held as the value of the SHAREd variable !*MODE and is thus accessible to the user in either mode.

RLISP itself is usually implemented in SLISP by writing the RLISP parser directly in SLISP, in order to "bootstrap" the full RLISP system. SLISP may in turn be implemented in a Lisp native to the actual machine. For example, on an Atari the native Lisp is Cambridge Lisp and on a VMS VAX it is usually PSL. The structure is

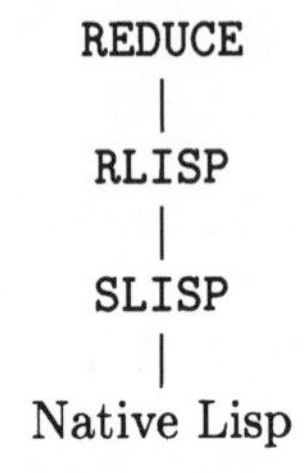

One by-product of the SLISP idea is that it is easier to make a given Lisp imitate SLISP than imitate Maclisp (or Common Lisp). REDUCE runs on a wider range of systems than MACSYMA mainly for this reason.

5.2.2 The Lisp programming model: Lisp data structures

The basic concept of Lisp's data representation (which may, but need not, be the way it is implemented on given hardware) is that of pairs of pointers, each pair being called a cell. The simplest object using this structure is a pair which can be presented diagrammatically as

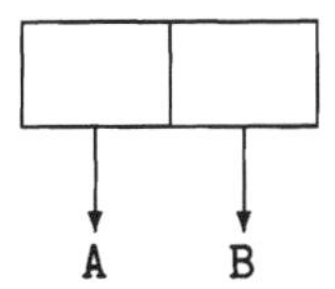

At the user level of Lisp such a structure would (in the simplest case) be printed out as `(A . B)`, and is called a *dotted pair*. The design of the implementation affects the maximum usable store. For example, in Cambridge Lisp a pair is in fact 8 bytes, giving 4 bytes for each half, of which the first byte is a tag giving the type of the quantity, and the other 3 bytes are the address of the item pointed at. Thus Cambridge Lisp can address $2^{24} = 16$ Mega-cells. Use of tags enables numbers and other data types to be recognized efficiently.

A *list* is a series of dotted pairs whose first element is an entry in the list and whose second element is a pointer to the next pair, with a last pair whose second half points to `NIL`. (It is important to note that Lisp lists are *not* exactly the same as the lists of algebraic mode REDUCE[1], although, for obvious reasons, their structure and the operations on them have similar names and effects.) A list is printed (naturally!) as a list of its entries (except the final `NIL`) surrounded by brackets, e.g. `(A B)` is the list diagrammatically presented by

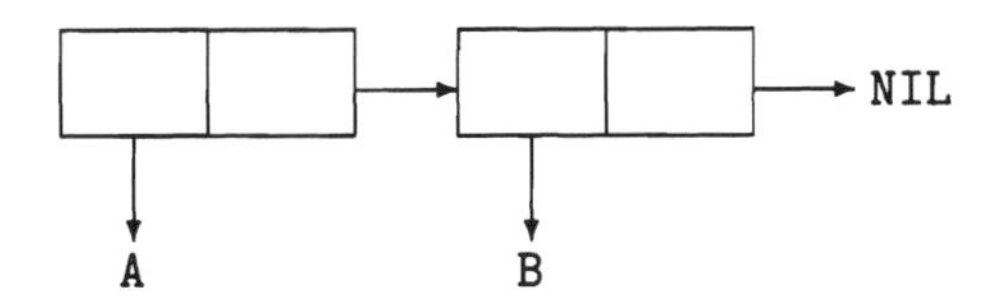

The ending convention is often implemented by internally representing `NIL` by a pointer to address 0 (or by some other standard, meaningless and easily tested address).

The elements of lists may themselves be lists, e.g. `((C D) B)` corresponds to the structure

[1]See §9.1.2.

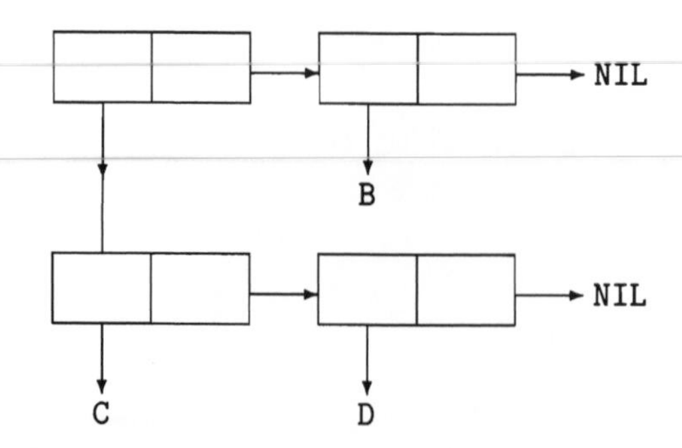

Any SLISP data which is not a (list or) dotted pair, i.e. is indivisible, is called an atom: `A`, 1, 7.0132×10^{-31} are atoms. Numbers are atoms but several different types (integer, floating point, and so on) are recognized. Most atoms which appear in a list are identifiers, i.e. variable or function names. There are two special atoms: `NIL` and `T`, which represent false and true, and are for this reason protected from being altered (which has the irritating consequence that `T` or `t` cannot safely be used as a time variable in REDUCE). `NIL` is (in the Lisps we use) also equal to `()`, the empty list: this breaking of the general definition that atoms are not lists is made for efficiency reasons (though, for example, in certain circumstances Cambridge Lisp will react to an attempt to treat `NIL` as a dotted pair as though it were an error). Atoms and lists together are S-expressions (symbolic expressions). Identifiers may have values (in which case they are called bound) and properties, as we describe later.

Because of the need for space, almost all Lisp systems have *garbage collection* in which cells no longer used are reclaimed for future use. To see the messages about this in Cambridge Lisp you can use the command

```
LISP VERBOS 1;
```

If your machine has limited memory this command (or its equivalent, if any, in your native Lisp) is worth using so that you can understand why a program appears slow when it garbage collects, and stop it if it is doing excessive amounts of garbage collecting.

Lisp usually has two other types of atomic data structure: strings and vectors. Strings are just sequences of characters (usually printable rather than control characters), and are input in a form such as

```
"This is a string"
```

in which the double-quote marks (*not* pairs of single quotes) are essential. These are exactly the same as REDUCE strings (though it makes more sense to say REDUCE strings are exactly Lisp strings). Vectors are like Lisp cells but with a number of pointers which need not be 2. A vector which is symbolically like

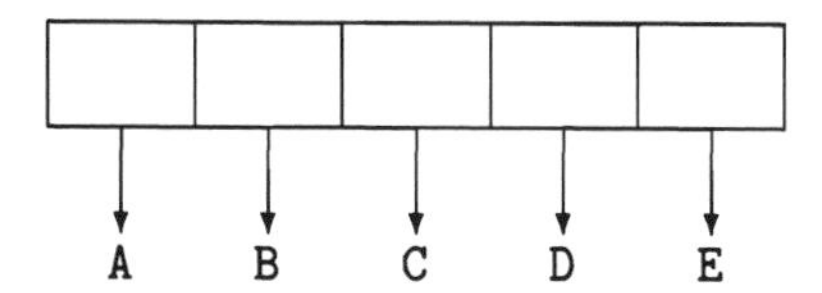

would be printed out as [A,B,C,D,E]. Lisp vectors are similar to arrays in other languages, and their elements are accessed by an index or offset. Remember that a Lisp vector is an atom; errors have been made because a user assumed that a vector would not satisfy the test for atoms.

An identifier (but not an atom which is not an identifier) may have the following associated with it: a value, a function definition, and a property list. In Standard Lisp (and some others)[2] an identifier cannot have both a value and a function definition. One side-effect is that you cannot safely use as variable names, even inside procedures, identifiers used as names of functions. Some identifiers in REDUCE are marked as "reserved" and listed in an appendix of the REDUCE manual: use of these may lead to an error message. Unfortunately the list does not include all the identifiers it is dangerous to redefine: for example, as explained in §4.7.1, **MAXDEG** should not be used as a variable because it is an important internal function in REDUCE.

The property list contains two types of entry: flags and properties. Flags are merely present or absent: for example, in REDUCE, reserved identifiers are marked with a flag **RESERVED**. In contrast, properties have values; properties and their values are usually stored on the property list as dotted pairs of the form

(propertyname . propertyvalue)

which becomes a list if the value is a list. As we shall see, almost all the REDUCE algebraic data is kept on property lists. In particular a variable, say x, which has (in algebraic mode) been assigned the value 2 does not have a Lisp value 2 (in general): instead that algebraic value is held on the property list of the identifier x, in the form shown at the start of the next section.

Identifiers may or may not be on the OBLIST, which is the name of the structure (not necessarily implemented as a list, and probably

[2]E.g. PSL allows this, but CULisp does not.

accessible only via special functions) containing the objects recognized by the Lisp reading functions. There is no reasonable way for the user to access or change an identifier not on the OBLIST. Such identifiers may be created, for example, as names of internal functions called only by other functions, not directly by the user: in this case, the internal function is put on the OBLIST while the functions which call it are compiled but then it is removed so that it is hidden from the user. To explicitly put an identifier on the OBLIST one uses `INTERN`, and to remove one, `REMOB`. (This provides a rather extreme way of avoiding identifier conflicts.)

5.3 The use of Lisp data structures for algebraic quantities

Global scalar algebraic variables in REDUCE have their values kept on their property lists under the property `AVALUE`. If we start REDUCE 3.3 and give as the first command

```
X := 2;
```

we would find that the property list of x would then (in Cambridge Lisp) read

```
((avalue scalar 2) used!*)
```

The property list of x here has 2 elements, the first being the property `AVALUE` with value[3] `(SCALAR 2)` and the second being the flag `USED!*`. Note that the integer 2 is held as itself (because numbers evaluate to themselves): numbers in other domains, such as rational, floating point or complex numbers, would be held in a different form which we describe later in this section. The command which enables one to read the property list will depend on the Lisp used. In RLISP based on Cambridge Lisp it would be

```
PLIST 'X;
```

where the quote mark (implying the `QUOTE` function) is essential. (As explained in the previous section, one must in general preface symbolic commands used in algebraic mode by the word `LISP` or `SYMBOLIC`, but the use of ' on the first argument implies that the mode of the command is symbolic.) Some other Lisps, in particular Portable Standard Lisp (PSL), use `PROP` instead of `PLIST`. With a long or complicated property list it may improve readability to apply the "prettyprinter",

[3]The precise form in which algebraic values are held under the `AVALUE` property depends on the REDUCE version and is liable to change.

which aligns and indents Lisp lists. (In many implementations this is supplied by the underlying Lisp, but REDUCE also has one for use when the Lisp does not provide such a facility.) To use the prettyprinter the required command would be (from algebraic mode)

```
LISP PRETTYPRINT PLIST 'X;
```

which writes its result in Lisp syntax. There is a similar function `RPRINT` which writes the result in RLISP syntax, but this is only for use on function definitions and commands, not on data structures, such as a property list, which have no RLISP form (i.e. where RLISP uses the Lisp form for such structures).

Now we can look at the way REDUCE stores different types of algebraic structure. Structures other than scalar values are mostly stored[4] under the property `RVALUE`, and the type of the structure is stored as the value of the property `RTYPE`. A matrix is stored as a list of rows in which each row is itself a list of the elements it contains: this corresponds closely with the form in which the data is input using the `MAT` function. For example:

```
Y := MAT((1,2),(3,4))$

PLIST 'Y;

((rvalue mat (1 2) (3 4)) (rtype . matrix) used!*)
```

Further discussion of this point can be found in §9.2.

One-dimensional arrays are held as Lisp vectors, whose length is determined by the size of the array; two-dimensional arrays are Lisp vectors whose elements are themselves Lisp vectors, and so on. The dimensions of an array are also stored explicitly on the property list of the array's name. REDUCE lists are just lists of the algebraic values of their entries; their storage is rather like that of the matrix above but `matrix` and `mat` in the above example are replaced by `list`. The stored value of an algebraic list $\{a,\ b\}$ does not contain the braces, and the entries, e.g. a, are stored not as atoms but in "pseudo-prefix standard quotient" form (see two paragraphs below for an explanation of this form). In fact, this form is used for the scalar values of the entries in other algebraic structures as well (e.g. for matrices and arrays).

Many other properties and flags are used by REDUCE as ways to store information about identifiers: for instance, a `LET` rule for `OP(X)`, where `OP` is an operator, puts entries on the property list of `OP` under the properties `KLIST` and `KVALUE`, in which the K stands for "kernel".

---

[4]In REDUCE 3.3: this is different in REDUCE 3.2 and earlier versions, and may change again in the future.

To round off this section we need to look at the way REDUCE stores a particular scalar algebraic value (remembering that matrices, arrays, and so on will involve more than one such value in general). In fact REDUCE has two standard internal formats for algebraic expressions, the "prefix form" which is a Lisp list whose structure we will explain when we have dealt with Lisp functions, and a "standard quotient" form, which is the form in which the final answer to a calculation is usually recorded. The standard quotient form provides a way to use internally a canonical form for rational functions in which polynomials are represented recursively. It can be depicted as in figure 5.2.

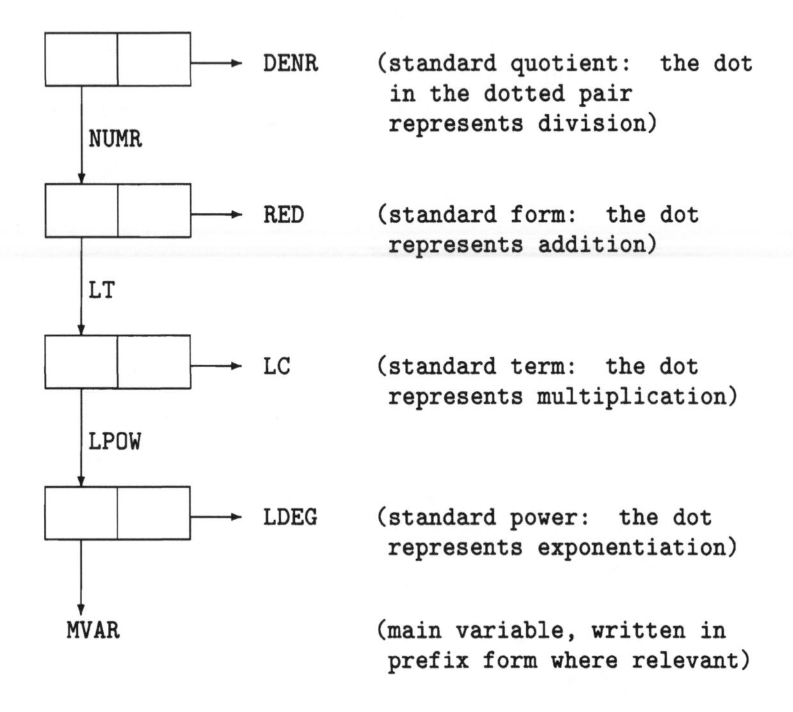

Fig. 5.2. The structure of a standard quotient: see the main text for a more detailed explanation

The meanings of the names in this diagram are explained in the following paragraphs. (See also the REDUCE 3.3 manual, §16.9.1.) The structure in Lisp is essentially the one described less formally in §2.6. The names given to the parts of the diagram are also the names of the internal symbolic mode functions used to select those parts from a standard quotient; the corresponding algebraic mode names were introduced in §2.6.

`NUMR` is the numerator and `DENR` the denominator; each has the same form (that of a polynomial). Since there can be only one numerator and one denominator the top Lisp cell in the diagram is the first cell of a list, the rest of the list being the terms of the denominator. A polynomial (e.g. the numerator or denominator) may be a number from the current number domain, which is just represented by itself, or it may not, in which case it will have the form shown in the lower part of the diagram. Such a form is called a standard form, so that a standard quotient is a dotted pair of standard forms, in which the dot signifies division.

If the number domain is not the integers, a number will be written internally as a structure like `(!:rn!: 1 . 2)` for the rational 1/2, `(!:ft!: . 0.5)` for its floating point equivalent, or `(!:gi!: 2 . 1)` for the "Gaussian integer" $2 + i$, and so on for complex floating and bigfloat numbers.

`LT` is the leading term of the standard form and `RED` the reductum; which term is considered leading is decided by the ordering of the variables in the polynomial. Only non-zero terms are given: the representation is sparse. Because a standard form may represent a polynomial with any number of terms, standard forms are lists of terms, and hence `NIL` must be considered an allowable standard form, since it is what we get when all the terms have been removed. So a standard form is `NIL`, or a number, or a dotted pair of a term and a standard form (the reductum). Each term is like the one shown in the diagram, and is called a standard term. In each dotted pair in the list of terms, the dot represents addition.

`LPOW` is the leading power and `LC` the leading coefficient. The coefficient will be a polynomial in all variables except the main one (i.e. the representation is recursive) and is therefore written as a standard form. The leading power is of the form shown, called a standard power. The Lisp cell must be a dotted pair, which will be a list if the coefficient is, and the dot represents multiplication.

Finally, in an individual power, `MVAR` represents the main variable and `LDEG` the degree: this form is only used for integer powers, so `LDEG` is a positive integer and the Lisp cell is a dotted pair, but not a list. Here the dot represents exponentiation.

The `MVAR` itself is a REDUCE kernel: in fact, as mentioned in §2.9, kernels are exactly those things permitted in this place in a standard power, and in normal circumstances are therefore exactly those things which are used in algebraic expressions but which cannot be broken down into standard power, term, form or quotient format: the exception is that turning off the switch EXP forces REDUCE to allow expressions that can be broken down in this way to nevertheless be considered as kernels if they are input as quantities raised to a power other than 1. A kernel which is not just an identifier is written internally as a Lisp list (a REDUCE prefix expression).

The standard quotient itself, when stored as the algebraic value of an identifier (e.g. under the property AVALUE) will be wrapped up as the middle element of a 3-element list whose first element will be the identifier !*SQ, which acts as an indication that the rest contains a standard quotient (rather than the prefix representation described in §6.4), and whose last element is a flag indicating whether the expression has been fully simplified. We shall refer to this "wrapped-up" standard quotient form as a *pseudo-prefix* form, because most internal REDUCE functions accept it as an alternative to a true prefix form.

Here is an example

```
ANS:=(3*A**2+2*B**3)/(A-B**2)$

PLIST 'ANS;

((avalue scalar (!*sq ((((a . 2) . 3) ((b . 3) . 2))
 ((a . 1) . 1) ((b . 2) . -1)) t)))
```

As discussed in §2.10, the specific form of the standard quotient for a given algebraic value can be altered in various ways. The most obvious is when the ordering of the kernels is altered (by the KORDER command). The other possibilities result from the use of switches, the EXP switch having already been explained. The switches which alter the current domain of numbers clearly will affect the definition of a standard form (by changing the meaning of the statement that a standard form can be a number), so COMPLEX, CONVERT, FLOAT, BIGFLOAT, MODULAR, NUMVAL and RATIONAL will affect this. The switches which affect the reduction of sums and fractions to lowest terms (see §2.10) will also have an effect: these are GCD, LCM and MCD. The FACTOR switch causes a factored rather than expanded form to be used for polynomials. There are also some more rarely used switches (see §2.10) which affect the internal (and printed) form, such as INTSTR, NOLNR, PRECISE, RATIONALIZE and REDUCED (and there are internal switches which affect the form at an intermediate stage of simplification). [NOLNR, the only one of these not introduced in §2.10, affects the form of the symbolic integral returned when INT encounters an integrand it cannot integrate, for example $x + (e^x/x)$. With NOLNR on, INT will not integrate even those parts which it could integrate, instead returning all terms as symbolic integrals.]

5.4 Further remarks on representations and simplification

The comments made earlier about representation glossed over a very important point in practice, which is that while the problems associated with representations of multinomials in independent variables (though not their solutions) are clear enough, one very soon has to consider whether the variables in the multinomials are really independent. When they are not, or worse still when one cannot decide whether they are or not, one faces some very difficult problems, some of which have even been shown to be insoluble (Buchberger and Loos 1983).

An element x is said to be *algebraic* over R if it obeys a polynomial equation, $q(x) = 0$, i.e. some element of $R[x]$ is zero. For the cases we shall be concerned with, R will be a field, E say, and any polynomial $q(x)$ such that $q(x) = 0$ will be divisible by the monic polynomial $p(x)$ of smallest degree n such that $p(x) = 0$. We use $(p(x))$ to denote the set of all elements divisible by $p(x)$. An obvious computational rule is to take the highest (n^{th}) degree term in $p(x)$ and substitute it wherever possible by the remaining terms. The resulting ring, which is denoted by $E[x]/(p(x))$, can also be considered as a vector space with a basis consisting of $1,\ x,\ x^2,\ \ldots,\ x^{(n-1)}$ and components with values in E. This ring will in fact be a field, called an algebraic extension of E.

A *transcendental* quantity over a ring R is one which is not algebraic over R. (That one cannot "square the circle" arises because π is transcendental over the rational numbers $\mathbf{Q}$.) The extensions of fields introduced earlier, in which x was assumed not to obey any polynomial equation, are then called transcendental extensions.

Starting from a polynomial ring with more than one unknown, one could define an algebraic extension using more than one, possibly multivariate, polynomial. In handling such extensions one can make use of the fact that for sets of polynomial equations, even if they are multivariate, there is a method (Buchberger's algorithm) for producing a minimal list of polynomials which must be zero (a Gröbner basis). The REDUCE package for this is discussed in §8.3.

Moreover one can build a *tower* of extensions, i.e. extend a field E to $E(x,\ \theta_1,\ \theta_2,\ \ldots,\ \theta_n)$ where each θ_i gives an extension (algebraic or transcendental) of the previous field $E(x,\ \theta_1, \ldots,\ \theta_{i-1})$. Unfortunately this idea is not without difficulties. For example, if a field is extended by any (finite) number of algebraic extensions, the result can be written as a single algebraic extension. (In the case where $\mathbf{Q}$ is algebraically extended, the result is called a field of algebraic numbers.) However, working in terms of this single algebraic extension makes expressions very complex and is not necessarily the best way: some more on this problem can be found in Davenport *et al.* (1988). The groups at Bath and Grenoble have each implemented large packages for handling algebraic numbers in REDUCE, and a third, written by E. Schrüfer, is distributed with REDUCE.

Algebraic numbers provide an example of the usefulness of *normal representations.* These are representations which are not canonical, in that quantities with the same value are not guaranteed to be represented in exactly the same way, but they are guaranteed to be represented as 0 if they equal 0. (For decideable sets of terms with a "minus" operation the existence of a normal simplifier implies that of a canonical simplifier, and there are interesting cases where this applies: see Buchberger and Loos (1983).)

Transcendental extensions can give even worse problems. For practical calculation, the problems already arise with functions of a single variable: for example, we know that $\sin x$ is transcendental over the field of rational functions $\mathbb{Z}(x)$, but $\cos x$ is not transcendental over $\mathbb{Z}(x, \sin x)$ since $(\sin x)^2 + (\cos x)^2 = 1$. The problems are in fact of two types: one which is normally assumed (but has never been proved) to present no difficulty and one which is insoluble. The first of these is the "constants" problem which is to check the relations between apparently different transcendental extensions (for functions, as distinct from constants, this can be checked using the "structure theorems": see Davenport *et al.* 1988 and §7.2). For instance, is $\pi + e$ a transcendental number over $\mathbb{Z}$? The second problem is that for certain classes of function, e.g. the class of functions built using the sin and absolute value functions on a ring $\mathbb{Q}(\pi, x)$, no algorithm reducing all functions in the class to a normal representation is possible (Buchberger and Loos 1983).

5.5 Exercises

1. Under what circumstances will REDUCE print a polynomial in recursive form, and under what circumstances will it print in distributed form?

2. How many terms would be needed to write $1 + x^5y^7$ in a dense representation with increasing total degree and x taken to be ordered before y?

3. In a REDUCE session, assign different types of algebraic entity (e.g. matrices, vectors, operators, procedures ...) to some variables and inspect the resulting property lists. What do you learn?

4. If

```
((((a . 2) . 3) ((b . 3) . 2)) ((a . 1) . 1)
 ((b . 2) . -1))
```

is a REDUCE standard quotient, what algebraic value does it represent, and to what Lisp prefix form does it correspond? [The second part may have to wait until after chapter 6.]

5. Make as much sense as you can of the property lists of the identifiers `EXP` and `SIN`. [You will find the prettyprinter useful for this exercise.]

6. Inspect the results of the following sequence of commands (replacing `PLIST`, if necessary, by the appropriate native Lisp command for inspecting property lists, probably `PROP`)

```
Y := EXP LOG X;
PLIST 'Y;
FOR ALL X LET EXP(LOG(X)) = X;
Y;
```

Use your observations to explain why the let-rule substitution fails in its obvious intention, and give one which will work.

7. Why was it safe to use `MaxDeg` as an argument name in the procedure in chapter 4, exercise 3?

6 Programming in (R)Lisp

6.1 The form of Lisp commands

In this very brief introduction to Lisp, we will try to avoid features very specific to CULisp (Cambridge Lisp), which we ourselves use, but we will sometimes mention differences between Lisps. There are many good (and some bad) books on Lisp: one we have found useful is Winston and Horn (1988) though this is based on Common Lisp (older editions used Maclisp). Cambridge Lisp's built-in commands are in lower case, though names in either case are allowed (and there is in fact an upper case version also). However, we will follow our usual conventions and put Lisp commands in upper case here for clarity.

If you want to use REDUCE's underlying Lisp directly, whichever it is for the implementation you have, you can enter it by

```
END;
```

and return to (i.e. restart) REDUCE by

```
(BEGIN)
```

A Lisp command consists of an S-expression (an atom or a list). Thus Lisp has the special property that commands and data are of the same form. If an atom is given, the value of that atom is returned. (Note: numbers always evaluate to themselves.) For example, in REDUCE, try

```
SYMBOLIC;              % or LISP;
INPUTBUFLIS!*;
```

to see all previous inputs (in Lisp form); the value of `INPUTBUFLIS!*` is a list of all the previous commands, and is where the `INPUT` command of REDUCE looks in order to re-use those old commands. (The similar list used by the `WS` command is called `RESULTBUFLIS!*`.)

If a list is given, say

```
(A B C)
```

Lisp evaluates this by applying the function A to the arguments B and C, i.e. Lisp assumes the first entry in a list is the name of a function, and responds with the value obtained by applying it to the rest of the list (the value "returned" by the function with the rest of the list as its arguments). If the command was given interactively (as distinct from being a call within a program) this value is usually shown at the terminal. Lisp is named after this basic property of applying a read-evaluate-print loop to lists: the word Lisp is short for list processing. In Cambridge Lisp the input is (by default) prompted by

```
Input:
```

and the output is signalled by

```
Value:
```

but other Lisps have other prompts (or none at all). The prompt in RLISP is the same as in REDUCE.

In RLISP the input format is apparently not in a list form, but in fact the first thing REDUCE does is to convert the input to Lisp form. In RLISP `(A B C)` is input as

```
A(B,C);
```

The switch `DEFN` (see §6.10 below) of REDUCE allows one to see exactly the lists thus produced. REDUCE recognizes that a list rather than an atom has been input by various means, one of which is whether there are arguments. It is for this reason that a function of no arguments, say `JUNK`, must be input to REDUCE as

```
JUNK();
```

and not as

```
JUNK;
```

since REDUCE reads the latter as an instruction to find the value of the atom `JUNK` rather than to evaluate the function whose name is `JUNK`.

If a list of lists is given, say

```
(A (B C) (D E))
```

Lisp evaluates the function B with argument C and the function D with argument E and then applies the function A to the values returned by the first (innermost) evaluations.

Having got this far we have to find out what some specific Lisp functions do. From the description of Lisp data structures in the previous chapter it is clear that the most fundamental commands must be those which pick out the two parts of a Lisp cell or form a new cell from two parts. These commands are called CAR, CDR and CONS. CAR and CDR are so named for historical reasons to do with the assembly language of the first machines on which Lisp was used (almost all other Lisp functions have mnemonic names): they each take one argument, a dotted pair, and respectively return the first and second halves. These are not entirely symmetrical if the cell is part of a list, since the CDR will then be a list but the CAR will not in general. CONS (so named, cf. §1.12, because it constructs: in Lisp it constructs a new cell) takes two arguments, and forms a dotted pair from them: it is the only function required to create new Lisp list structures since all the others can be defined in terms of CONS.

QUOTE, which just returns its argument without alteration, is a frequently used function: it can be written as ', so that

```
'B;
```

is the same as

```
QUOTE(B);
```

Try

```
'INPUTBUFLIS!*;
```

in RLISP and see the difference from the earlier example. QUOTE is very useful in coping with one of the most frequent problems in Lisp which is that many functions, including CAR, CDR and CONS, work not on the arguments as given but on the values of those arguments.

It is worth noting that CAR works on a Lisp list much as FIRST works on a REDUCE list, while CONS has as its analogue (and as its synonym in symbolic mode) the infix operator "." and CDR corresponds to REST applied to REDUCE lists. Remembering the properties of QUOTE, CAR, CDR and CONS (or equivalently ".") we see that

```
CONS('A,'(B C));          %      gives (A B C)
'(A B) . '(C);            %      gives ((A B) C)
```

while if L has a value which is a dotted pair

```
CAR(L) . CDR(L);
```

will return the value of L itself, whereas

```
CAR('L);
```

and

```
CDR('L);
```

will be errors since the identifier L is not a dotted pair (though its value is).

We now return to the question of how Lisp functions deal with their arguments. Functions have two important types of difference in behaviour:

1. They may either evaluate their arguments, i.e. work on the values, or they may not. These are called respectively EVAL and NOEVAL functions.

2. They may either take a fixed number of arguments, or they may consider their argument to be a single list of arbitrary length. These are respectively called SPREAD and NOSPREAD functions.

Most Lisps have EVAL, SPREAD functions (often called EXPRs) and NOEVAL, NOSPREAD functions (often called FEXPRs). In some Lisps compiled EXPRs are called SUBRs, to help identify the compiled forms of the functions, and similarly compiled FEXPRs are called FSUBRs. Cambridge Lisp also has the other two possibilities, but these are not used by REDUCE, which in fact also avoids FEXPRs[1]. Most of the elementary Lisp functions (and most REDUCE functions) are EXPRs: when required, REDUCE can achieve the effect of an algebraic FEXPR by its parsing routines which can collect the arguments supplied and form a list from them without algebraic evaluation (see §9.1 for more on this point).

Most Lisps have functions called `CAAR`, `CDDR`, `CDAR` etc, which are combinations of repeated `CAR` and `CDR` operations [in left-to-right order, e.g. `CADR(L) = CAR(CDR(L))`]. In fact the selectors for the parts of a standard quotient, such as `LC`, are just defined directly as appropriate ones of these combinations, `LC` of a standard form being `CDAR`, for instance.

To set a variable one can use `SET`, e.g.

```
SET('TEMP,100);
```

As writing the ' gets painful, and is easily omitted by accident, there is also the more-commonly used `SETQ` (meaning "set quote"), e.g.

```
SETQ(TEMP,100);
```

has the same effect. The value returned is the value set, i.e. here 100. The infix operator ":=" is equivalent to `SETQ` when applied in REDUCE

[1]The use of FEXPRs is discouraged by Anthony Hearn because they are very difficult to compile efficiently.

symbolic mode: in algebraic mode ":=" has another function SETK, named from "set kernel", as its main meaning.

As described in §4.7.1 some variables are GLOBAL, i.e. it is an error if you declare them as local variables in a block statement or use them as arguments in a procedure. Arguments used only in a compiled function are LOCAL (and are usually inaccessible by name from outside the function). Most variables are FLUID: in this case if they are used locally, the Lisp system stacks the values when it enters the block or function, and resets SCALAR variables to NIL, which in algebraic mode is equivalent to assigning these variables the initial value zero, see §3.7.5, whereas INTEGER variables are explicitly set to 0. On exiting the function the original values are restored. The precise mechanisms for doing this vary widely between different Lisps and are an interesting area of debate about efficiency and logicality of structure.

SCALAR variables in algebraic mode procedures are not handled like algebraic variables, i.e. it is their Lisp values which are used not their AVALUE properties: this accounts for the problem, discussed in §4.7.2, in passing their values implicitly as part of the environment to other algebraic procedures. It is worth repeating, as stated in §4.7.1, that a variable declared SCALAR in a block, unless explicitly declared to be FLUID, will be treated as FLUID if interpreted but LOCAL in a procedure that is compiled (in most implementations): hence use of (for example) MAXDEG as a variable name may work fine if the procedure is compiled but not if it is interpreted – which can lead to considerable bafflement when trying to debug a program!

6.2 More functions on lists

There are many more Lisp functions than the few mentioned so far. As well as CONS, one has APPEND which applies to a pair of lists, e.g.

```
APPEND('(A B),'(C D));          %      gives (A B C D)
```

and LIST, e.g.

```
LIST('(A B),'(C D), 'E);        %      gives ((A B) (C D) E).
```

The meanings of the further useful functions on lists LENGTH and REVERSE are obvious (note that APPEND, REVERSE and LENGTH also work for REDUCE algebraic lists, as described in §1.12).

Most functions build new lists, e.g. APPEND and REVERSE do, but some do not, e.g.

```
NCONC('(A B),'(C D));
```

returns the same value as

```
APPEND('(A B),'(C D));
```

but NCONC replaces the NIL after the B by a pointer to the first cell of the second list (the name NCONC means NIL concatenate) while APPEND first copies the (A B) list to new places and then changes the NIL on this new list. To see the difference, try the following sequence of commands (in symbolic mode)

```
L := '(A B);
M := '(C D);
```

and then either

```
APPEND(L,M);
```

or

```
NCONC(L,M);
```

followed by

```
L;
```

NCONC can thus be described as a destructive form of APPEND, since it destroys the first of its two arguments. When Lisp experts are discussing programs they may ask whether a function is destructive or copies its arguments. RPLACA and RPLACD replace (destructively) the CAR and CDR of a pair. Similarly most systems have a destructive REVERSE (REVERSIP – reverse in place – in RLISP and REVERSEWOC – reverse without copying – in CULisp).

A very common use of RPLACA in REDUCE is to implement the properties of switches. The last part of a pseudo-prefix form is T or NIL depending on whether further simplification is needed or not. This is reset by certain switches and functions (such as LET) in the following way. When a simplified standard quotient is constructed, the last part of the pseudo-prefix form is made to point to the value of the variable !*SQVAR!* which will be the list (T). When those switches or functions which might imply re-calculation are used, a function RMSUBS is called (cf. chapter 2, exercise 6) which does a RPLACA on the value of !*SQVAR!*, setting it to NIL, thus causing the last part of every pseudo-prefix form to become NIL, and then gives !*SQVAR!* a new value, in a new place, which is again (T), so that when a given pseudo-prefix form is re-simplified the final marker will be altered back to T. (Destructive functions are also used in the matrix handling code – see §9.2.)

The function EVAL calls explicitly for an extra evaluation, e.g. if A has value B and B has value C,

```
A;
```

returns B, but

```
EVAL(A);
```

returns C.

All Lisp functions, without exception, return something. (This is not apparent in the algebraic mode of REDUCE, simply because many functions, including all those described earlier as returning nothing, in fact return `NIL`, and the REDUCE printing routines ignore this as uninteresting in algebraic mode.) However, their real purpose may be to achieve some side-effect. For example `SETQ` returns the value set but its main purpose (usually) is the side-effect that a value is assigned to its first argument. Functions whose whole purpose is the side-effect often return simply `NIL`. There are arguments about Lisp programming style. Many people (including Anthony Hearn, the author of REDUCE) prefer the "functional programming" style in which the purpose of a function is what it returns rather than its side-effects: although it is often inconvenient, and sometimes very hard, to maintain this style in all circumstances, contributors to the REDUCE program library therefore try to write in this style.

6.3 Yet more basic RLISP functions

As the numbers of arguments of the following functions, and their side-effects and values returned, are mostly fairly obvious, we do not give full details of each one. They are divided into classes with different purposes or types of argument.

Arithmetic

`ABS`, `ADD1`, `DIFFERENCE`, `DIVIDE` (gives quotient and remainder), `EXPT`, `MAX`, `MIN`, `MINUS`, `PLUS`, `REMAINDER`, `QUOTIENT`, `SUB1`, `TIMES`. (`ADD1` and `SUB1` act on integers and respectively add and subtract 1 from them.)

Number type translation

`FIX`, `FLOAT`.

Functions on lists

`MEMBER`, e.g. `MEMBER('A,L)` is true [and returns `(A B C)`] if L has the value `(A B C)` but not if L has the value `((A B) C)`. `MEMBER` has as its value the remainder of the list (or `NIL`). The test for equality uses `EQUAL` (see below).

MEMQ same but using EQ (see below).
SUBST('D,'A,L) substitutes D for A in the value of L: e.g. if L is (A B C) this would return (D B C).
It should be noted that whereas MEMBER and MEMQ inspect only the top level of a list, SUBST will work recursively on elements which are themselves lists: in fact, for SUBST, L can be any data structure, not necessarily a list.

Logical operators

Lisp has a large number of testing functions which in principle return true or false. In many systems (including RLISP) any non-NIL value is true, allowing a logical test, if true, to return a more interesting value than T. These test functions include, and can be combined with, OR, AND and NOT: e.g. in Lisp 1.6

> (OR (NUMBERP 1)) returns T, and
> (NUMBERP (OR 1)) is false (returns NIL)

whereas in CULisp

> (OR (NUMBERP 1)) returns 1
> (NUMBERP (OR 1)) is true and returns 1

Many, but not all, these tests (called predicates) have names ending in P. Some tests available in RLISP are:

tests of data type:
: ATOM, CODEP (is it binary code, i.e. a compiled function?), CONSTANTP, FLAGP (is the variable flagged? – requires the flag-name as second argument), FLUIDP, GLOBALP, IDP, KERNP (is it a kernel?), LISTP, PAIRP (is it a list or dotted pair?), STRINGP, VECTORP.

tests of properties of the values:
: DIGIT, FIXP (is it an integer?), FLOATP (is it a floating point number?), LITER (is it a literal, i.e. a letter?), NULL (is it NIL?), NUMBERP.

tests of arithmetic:
: EQN (numerical equality), EVENP, GREATERP, LESSP, MINUSP, ONEP, ZEROP.

tests of the operating environment:
: BATCHP, FILEP (these two functions may not be available or useful in all implementations), TERMINALP.

It is worth noting that a REDUCE algebraic operator with a name identical to one of the names listed above does not generally invoke the same function.

Equality can be tested by `EQ` or `EQUAL`. `EQ` compares pointers and so will not detect equality of two distinct copies of the same object, but is more efficient when it is applicable. `EQUAL`, which really tests equality of the contents of the two structures pointed to, is more reliable.

There is a family of six functions of the form `FOREACH` ... in RLISP, similar to those explained in §3.1.2. Each applies the statement given as the final argument to one of

(1) the car of the remaining list (RLISP keyword IN)
(2) the cdr of the remaining list (RLISP keyword ON)

and returns one of

(a) the list made by concatenating the results, which are lists (JOIN)[2],
(b) the list whose elements are the various results (COLLECT),
(c) `NIL` (though output may be written by the functions called) (DO).

In most Lisps there are equivalents which have names of the form MAPxxx, i.e. (MAPxxx *function list*), as shown in the following table:

	JOIN	COLLECT	DO
IN	MAPCAN	MAPCAR	MAPC
ON	MAPCON	MAPLIST	MAP

These Lisp equivalents are not directly used by REDUCE except in a few places.

We should also note the useful functions `EXPLODE` and `COMPRESS` which respectively pull an atom (other than a vector) apart into a list of its individual characters (each of which is `INTERN`ed), and make an atom, which is not `INTERN`ed and will not be a vector, from a list of single character identifiers – for an example of their use see the definition of `MKID` given in §6.5 below.

6.4 Prefix form and conversion between internal forms

The REDUCE prefix form is simply a Lisp list in which the functions are the ones of algebra (or arithmetic) such as `times`, `df`, and `quotient`. For example,

```
(a+b)/(c**3 + a*df(x,y))
```

would be

---

[2]In REDUCE 3.2 the JOIN keyword was called by the Lisp name CONC.

```
(quotient (plus a b) (plus (times (df x y) a) (expt c 3)))
```

In REDUCE documentation it is this which is called the "algebraic" form. Like the standard quotient form the actual details of the internal structure of a Lisp prefix form are affected by the state of some of the REDUCE switches.

REDUCE provides RLISP functions to convert between prefix form and the standard quotient form. For example, to convert from the prefix form to standard quotients there are MKSQ, MK!*SQ (which adds the !*SQ part), MKSP (standard power), MKSP!*, MKSFPF ... while for conversion of a standard quotient or standard form to prefix form one can call PREPSQ or PREPF: e.g.

```
PREPSQ CADR X;
```

returns the true prefix form of the algebraic value of X assuming that X is a pseudo-prefix expression.

There are also functions of the form !*u2v where u and v may be appropriate ones of:

A algebraic (prefix form)
F standard form
K kernel
P standard power
Q standard quotient
T standard term

Note that

```
SHARE X;
```

sets the Lisp value of X to be the same (even after calls to the simplifier or other functions) as its AVALUE, with the consequence that CLEAR (which usually clears the AVALUE) will not work[3] on X. SHARE provides a way of taking algebraic values into symbolic mode, and vice versa.

6.5 Defining a new EXPR

This is done by PROCEDURE which (except if preceded by a declaration that a different function type is desired, see §6.8) internally becomes the function DE (for Define Expr; DEFUN in Maclisp), which has the Standard Lisp form

(DE *name* (*arguments*) (*action of function*))

[3]But see the footnote at the end of §4.7.2.

DE returns the name of the new function, but its purpose is to set up the associated function definition (this is the side-effect), e.g. the RLISP (REDUCE symbolic mode) definition

```
PROCEDURE F!-TO!-C TEMP;
 (TEMP - 32)/1.8;
```

is the same as

```
(DE F!-TO!-C (TEMP) (QUOTIENT (DIFFERENCE TEMP 32) 1.8))
```

and returns

```
F-TO-C
```

but of course thereafter one has a function F!-TO!-C which will convert Fahrenheit to Centigrade (Celsius) temperatures. To avoid ambiguity, it is a good idea to precede PROCEDURE by the intended mode.

Here is a really useful function (defined in the source file ALG1 and used in §9.3)

```
symbolic procedure mkid(x,y);
  % creates the ID XY from identifier X and
  % (evaluated) object Y.
  if not idp x then typerr(x,"MKID root")
   else if atom y and (idp y or fixp y and not minusp y)
    then intern compress nconc(explode x,explode y)
   else typerr(y,"MKID index");
```

Note that the combination INTERN COMPRESS creates a new identifier on the OBLIST from the list of characters created by the NCONC and the EXPLODEs.

A symbolic procedure can be made available in algebraic mode by declaring it with a statement

SYMBOLIC OPERATOR *ProcedureName*

which flags the procedure's name with the OPFN flag. An example was given in §3.5. The same flag is used for algebraic procedures, so the effect of the declaration is to mark a symbolic procedure as if it were algebraic. There will of course still be a difference in how the procedures act, i.e. whether the algebraic or symbolic evaluation rules are used.

6.6 Control structures

IF *test1* THEN *action1*
  ELSE IF *test2* THEN *action2*
      ...

is equivalent to the Lisp function

```
(COND (test1 action1)
      (test2 action2)
      ...           )
```

This evaluates each *test* (which must be a single S-expression) in sequence, and for the first one that evaluates true does the action and returns its value as the value of the statement. If all *tests* fail IF (or COND) returns NIL. Note that the side-effects of any *test* that is evaluated will be carried out and remain in force even if that *test* evaluates as false.

Block structures are important, and in RLISP, as in algebraic mode REDUCE, these are supplied by the BEGIN and END keywords. The BEGIN/END block becomes a Lisp PROG function. It takes the form

(PROG (*variables*) *S-expressions*)

The variables are by default fluid when interpreted, so any external values are stacked, and local when compiled (see §6.1); in the latter case, the name is only used in the input program, not in the compiled code, so no stacking is required or carried out. The RLISP declaration of these variables is the SCALAR declaration (the INTEGER declaration has similar effect but initializes the value to 0 rather than NIL). The S-expressions are evaluated in sequence. If one of the S-expressions is an atom, it is treated as a label, and the Lisp command (GO *label*) does the obvious thing (the colon terminating the label in the REDUCE or RLISP input is removed for Lisp). An RLISP block may explicitly be told what to return, by RETURN *expression*, just as in algebraic mode, but otherwise returns NIL. In Lisp, a variant on PROG is PROGN (named from PROG NIL) which has the last evaluated expression as its value: this is what a REDUCE or RLISP group statement created by the symbols "<<" and ">>" translates to in Lisp. Lisp purists prefer (when it is reasonable to do so) to achieve the effects of FOR, DO and GOTO by recursion, rather than use the GO and RETURN features inside a PROG (which are regarded as supplied for FORTRAN programmers who cannot lose their bad habits) but still find the block structure of BEGIN/END (i.e. PROG) useful.

The other control structures of REDUCE are defined as Lisp MACROs; see §6.8 where the example of WHILE is given.

6.7 More on manipulating Lisp data

The property list of an identifier is accessed in RLISP by

```
PUT('identifier, 'propertyname, 'propertyvalue)
GET('identifier, 'propertyname)
```

which respectively set and find the value. Note that PUT and GET are EXPRs. The identifier and property name arguments to PUT and GET are often given explicitly rather than as the values of other expressions and we have therefore shown these arguments in quoted form above (to emphasize the need for the use of QUOTE in such cases). For example,

```
GET('X, 'AVALUE);
```

returns the AVALUE of X.

A flag can be respectively set and tested by

```
FLAG('(list of identifiers to flag), 'flagname)
FLAGP('identifier, 'flagname)
```

Property lists in some Lisps include the value and the printname of the atom, and some Lisps allow an identifier to have both a value and a function definition. None of these apply to CULisp, and the latter is specifically forbidden in the Standard Lisp definition, although PSL allows it. The correct and portable way to put or get a function definition is to use the Standard Lisp function PUTD (usually via DE etc.) or GETD. Use of these is illustrated at the end of §D in the Appendix. (However, in Lisps such as CULisp, function definitions can also be manipulated like variable values.)

Properties, flags and function definitions can be removed by using respectively

```
REMPROP('identifier, 'propertyname)
REMFLAG('(list of identifiers to unflag), 'flagname)
REMD('functionname)
```

Note that REMFLAG matches FLAG by taking a *list* of identifiers as its first argument.

A-lists (association lists) are lists of a particular form, namely with elements which are dotted pairs:

```
((key1 . value1) (key2 . value2) ... )
```

These provide a useful data (sub-)type which, for example, gives ways of stacking values, e.g. by an a-list for each function at each call whose entries are the stacked variable names (keys) and values, or an a-list for each variable whose keys are the functions doing the stacking. The function

```
ASSOC(key, a-list)
```

returns the pair with the key as first element (or NIL). REDUCE uses a-lists to store the inputs, the outputs, and most substitutions arising from LET or MATCH commands.

Finally we should mention vector[4] manipulation by the commands MKVECT, PUTV, GETV, UPBV (upper bound on the index of the vector).

6.8 FEXPRs, LAMBDA expressions and MACROs

To define FEXPRs in SLISP we use DF (like DE), which obviously has nothing at all to do with the algebraic differentiation operator. This is not used much in RLISP, and REDUCE itself does not use FEXPRs (see §6.1). However, a FEXPR can be defined by starting a procedure definition (cf. §4.1.1) with

```
SYMBOLIC FEXPR PROCEDURE ...
```

One can write a function without giving it a name, by using LAMBDA (meaning "for all") e.g.

```
(LAMBDA X; IF CDR X=0 THEN CAR X ELSE CAR X+1)
  DIVIDE(M,N);
```

is the definition of the least integer greater than m/n (for $m, n > 0$) used in the REDUCE source: the LAMBDA expression in the parentheses is applied to the result of the call of DIVIDE. Note the syntax of a LAMBDA expression in RLISP, which is a little unusual. LAMBDA is mainly used when one wants to put such a definition inside a function which will apply the new definition. One such function is APPLY, e.g. in Cambridge Lisp

```
L := '(1 2 3);
APPLY('PLUS,L);
```

overcomes the problem that PLUS(L) would fail here[5]. Another example of the use of LAMBDA is the way WHERE[6] is implemented: by turning DEFN on (see §6.10 below) one can discover that in algebraic mode

---

[4]In Maclisp and its derivatives one has arrays instead.

[5]While this provides an example which is easy to understand, it too would fail in Standard Lisp, because in Standard Lisp APPLY can only be used on EXPRs and LAMBDA expressions and not on FEXPRs or MACROs, and in order to be usable on a list of arguments of arbitrary length, PLUS must be a MACRO or FEXPR. In Cambridge Lisp it is a FEXPR and Cambridge Lisp's APPLY works on FEXPRs, but not on MACROs.

[6]Introduced in §2.7.

```
X**2 WHERE X=U;
```

becomes

```
(aeval ((lambda (x) (list 'expt x 2)) 'u))
```

and similarly in symbolic mode

```
(IF CDR X=0 THEN CAR X ELSE CAR X+1) WHERE X=U;
```

becomes

```
((lambda (x) (cond ((equal (cdr x) 0) (car x))
 (t (plus (car x) 1)))) u)
```

Finally we come to the most difficult function type to handle, namely MACROs. What makes them difficult is that they first form an S-expression and then evaluate it, amounting to two steps of evaluation: they provide a splendid way to write unintelligible code! In RLISP they are defined by

SYMBOLIC MACRO PROCEDURE ...

which transforms to

(DM *name* (*parameter*) *body*)

Such a definition first uses the *body* to define an S-expression, using the list of arguments given to the MACRO as the value of *parameter* (without evaluating that list at this stage): this is the first evaluation. Then it EVALs the result. The argument list of a MACRO is assumed to include its *name* as the first entry on the list of variables, so that, for example, to give CAR a more mnemonic name one could use[7]

```
(DM FIRST (S) (CONS 'CAR (CDR S)))
```

In RLISP one would achieve this by

```
SYMBOLIC MACRO PROCEDURE FIRST S;
   CONS('CAR, CDR S);                  % or 'CAR . CDR S;
```

It is worth noting that REDUCE also provides some other sorts of MACRO, including in particular the SMACRO type which is close to the sort of macro one has in other programming languages such as C and assembler. We will not now go further into this, but there is an example in §9.1.2.

The MACRO idea is used to define the control structures that are not included in the Standard Lisp definition (including those added

[7]Without giving the details, we note that before adaptation for RLISP, CULisp macros are a little different. In fact, the form MACROs take varies considerably between Lisps.

in the "extended syntax" described in the appendix to the Standard Lisp report (Marti *et al.* 1979)). For example, the `WHILE` structure of REDUCE is implemented by

```
symbolic macro procedure while u;
   begin scalar body,bool,lab;
      bool := cadr u; body := caddr u;
      lab := gensym();
      return mkprog(nil,
               list(lab,list('cond,list(list('not,bool),
               list('return,nil))),body,list('go,lab)))
   end;
```

Note here again the useful function `GENSYM`[8] which generates a unique symbol and thus provides a way to have non-conflicting but unbound names inside procedures; in this case it is used for the label for the `PROG` constructed by the `MACRO`.

Note also that the whole of REDUCE is the Lisp function `BEGIN` which of course has very many other functions which it calls recursively.

6.9 Tracing and debugging

The `TRACE` function (and `UNTRACE`) are useful. They will show, whenever a traced Lisp function is called, what arguments it was given and what value it returned. The precise syntax of the `TRACE` and `UNTRACE` functions, and even their names (for which `TR` and `UNTR` are common variants), may vary from one Lisp to another, but most have some functions with the same purpose.

To be more specific, in both PSL and CULisp, `TRACE` and `UNTRACE` are EXPRs that accept one argument, which should evaluate to a Lisp list of identifiers. This list will almost always be supplied as a quoted list, in which case the quote will ensure that REDUCE processes the statement in symbolic mode, even if the default mode is algebraic. Hence, the following work correctly in either algebraic or symbolic mode, in REDUCE implemented using either PSL or CULisp:

`TRACE '(`*FunctionName1 FunctionName2* ... `)`
`UNTRACE '(`*FunctionName1 FunctionName2* ... `)`

(Because `TRACE` is a REDUCE algebraic-mode operator that returns the trace of a square matrix, it will never be possible to use it as a function-tracing command directly from algebraic mode, i.e. without something to force it to be interpreted as a symbolic-mode command.)

In CULisp, `TRACE` and `UNTRACE` also accept a single identifier as

[8]For its algebraic uses see §4.7.3 and §4.13.

their argument. In PSL, there are also MACRO functions called TR and UNTR that accept an arbitrary number of (unquoted) function identifiers that are not evaluated, and set or clear tracing on them. The REND source file includes some code that appears to be intended to make TR and UNTR available as commands like ON and IN, i.e. in both algebraic and symbolic modes they should accept either one or a comma-separated sequence of identifiers of functions to be traced or untraced, *which should not be quoted.* In PSL REDUCE this usage works, but produces rather confusing output and spurious error messages in algebraic mode, whereas in Cambridge REDUCE it really does not work properly. We anticipate that TR and UNTR will be the correct and documented ways to set and unset tracing in future versions of REDUCE but until the problems with them are corrected our advice is to use TRACE and UNTRACE as illustrated above.

In RLISP and REDUCE, a backtrace of errors is given by turning on the BACKTRACE switch. (In CULisp, one can control the amount of diagnostics produced by an error via the function BACKGAG(n), and information on file loading and garbage collection is given if the "verbosity" is reset to something above the default value of 0, e.g. by

```
lisp verbos 3;
```

There should be similar control facilities in other Lisps – for example, in PSL the switch GC, initially on (i.e. the global Lisp variable !*GC with initial value T), enables printing of garbage collection messages.)

All of these are useful tools for detecting what has gone wrong. There are also error reporting routines built into many of the functions of REDUCE; these can help to pick up simple errors like supplying the wrong numbers or types of arguments, or misplacing brackets (a task eased considerably, incidentally, by preparing input files using an editor such as a version of Emacs which shows how pairs of brackets are matched).

6.10 The structure of REDUCE and its additional modules

REDUCE consists of a basic system and a series of auxiliary programs which are loaded as the need arises (the dividing line between these two classes depends somewhat on the specific implementation). Moreover, it is usual in the source code to subdivide the programs into modules, each of which, after compilation, would be reasonably small and so load quite quickly when needed. The actual source code files therefore cannot, without confusion, be referred to as modules, although in many respects they behave similarly.

An implementation of REDUCE starts, if necessary, with some files which are specific to the Lisp system and enable the actual Lisp to read and correctly interpret SLISP in RLISP syntax. The source code for REDUCE itself is usually held in files with names of the form *name*.RED, but this may vary between systems, and we will omit any "extension" such as .RED from now on. We describe the source code files for the Atari version of REDUCE 3.3 in the next paragraph and, as an indication of the sizes of the various parts of REDUCE, give their sizes (these sizes, and the names, may be different in other REDUCE versions).

The files RLISP (100K) and REND (12K) are read to build RLISP itself. REND contains system specific definitions such as adaptations of the file handling functions to the operating system, and may therefore vary between implementations: other files should not vary. The main algebra features are contained in the source files ALG1 (131K) and ALG2 (137K). There are a number of much smaller auxiliary programs: RCREF (28K) contains the cross-referencer (which produces, from a source code file, a list of the functions it contains, their variables, and the functions they call); UTIL (39K) contains the input editor ED (internally referred to as CEDIT) and the prettyprinter (which may be commented out if the underlying Lisp provides one already, as is true in Cambridge Lisp); MATR (28K) is the matrix handler; and there are also small programs which may not be in all implementations, such as the mathematical library MATHLIB (15K) and a table of entry points to the main programs ENTRY (5K). The more substantial chunks of code are the SOLVE package (34K), T. Sasaki's big floating point package BFLOAT (91K), the integrator INT (160K), and above all the greatest common divisor and factorizer programs EZGCD (215K) and FACTOR (77K) which work together (and were both in one part of the source in REDUCE 3.2). The sizes of these programs give some idea of how hard they were to produce, and in particular illustrate that making certain calculations algorithmic, notably integration and factorization, is quite difficult. The integrator was written by Arthur C. Norman and P. Mary Ann Moore with later additions and amendments by Stephen J. Harrington, John P. Fitch and James H. Davenport, and the factorizer is by Norman and Moore again. These programs are discussed further in chapter 7.

All these are usually either loaded into the main image of a REDUCE implementation, or are compiled as separate modules in such a way that pointers to these modules are compiled into the basic image. There are many other pieces of source code which can be compiled either in the same way or as binary code without compiled-in pointers (and which then must be loaded explicitly by the user rather than automatically by the system) but are not included in the usual basic system. One file which we believe should be in this category but which is sometimes loaded (for example, in the versions we use) is the HEPHYS (24K) package (chapter 17 of the REDUCE 3.3 manual). For our own use, we often compile in packages for vector calculus or exterior calculus, ordinary differential equations, power series, etc.

When REDUCE is started, the function `BEGIN` is called. This tries to read an input statement using the reader defined in RLISP. If the first word of the statement has the `STAT` property, the value of that property is a function used to read and parse the rest of the statement. Having read it, the `FORMFN` properties may be used to re-parse it and if necessary it is converted between symbolic and algebraic mode parsings. Then the resulting (by now Lisp) statement is output if `DEFN` is on (unless the statement was `OFF DEFN` of course): otherwise the Lisp statement is evaluated and its value returned (and may be output, after appropriate formatting, depending, as usual, on the terminator chosen).

[More precisely, when the switch `DEFN` is on the following REDUCE commands are still executed, rather than being output in Lisp form: `ON`, `OFF`, `IN`, `OUT`, `SHUT`, `CONT` (but not `PAUSE`), `END`, `BYE`, `QUIT`, `TR`, `UNTR`, `ALGEBRAIC`, `SYMBOLIC` and `LISP`. This behaviour is mainly a consequence either of the keyword being flagged `IGNORE`, which is primarily for use during compilation, or of the keyword being handled specially by the parser.]

The additional packages distributed with REDUCE 3.3 included ALGINT (173K) for integrating algebraic functions, EXCALC (90K) for exterior calculus, GENTRAN (117K) for generating FORTRAN and C programs, GROEBNER (72K) for Gröbner basis calculations, and SPDE (36K) for finding symmetries of partial differential equations. These are discussed further in chapter 8 (and, in the case of EXCALC, in Dr. McCrea's lectures at the Rio school, which appear in volume 2 of this series). Since the launch of REDUCE 3.3 a networked library[9] has been introduced, which has a substantial and growing list of contributions including output formatting in TeX, vector calculus, truncated power series and so on.

If users wish to contribute to this library they are invited to do so, subject to some quality checks on documentation, examples and so on. It is worth noting that there is a certain feeling among the REDUCE community that for efficiency and in order to facilitate good programming style, extra programs should be at least mainly written in symbolic rather than algebraic mode. This is certainly the case

[9]See the Appendix, §F.

with most if not all of the library programs. However, we take a less dogmatic view; even at some cost in efficiency, there may be parts of programs which are much simpler to write (and hence to debug and maintain) in algebraic mode, so that this is the better course unless efficiency becomes paramount. As more facilities become available in algebraic mode (a process to which the examples in chapter 9 contribute some small steps), using purely algebraic mode will become easier. However, we do think it is good practice to preface each statement in a source file by its mode; then the user can load the file correctly regardless of the mode currently in use, and without problems restoring that mode afterwards[10].

6.11 The REDUCE simplification functions: an outline

The simplifications applied to a value depend on the data type of that value in REDUCE: an appropriate function is held as the `EVFN` property of the data type. What we describe next is for the scalar type, though other simplifiers generally call that recursively.

All the algebraic commands convert to calls to the internal functions `AEVAL` (which returns a standard quotient in pseudo-prefix form) and `REVAL` (which returns a true prefix form), and both of these, via `REVAL1` and `REVAL2`, the latter of which converts a standard quotient to the required return form (prefix or pseudo-prefix), call the simplifier function `SIMP!*` which calls `SIMP` (both of these return a standard quotient), which calls a network of a large number of functions (at least 120) which it would be impossible to describe fully here. The `SIMP` function maintains a list of previously simplified expressions (for obvious efficiency reasons) called `ALGLIST!*`, and a counter `SIMPCOUNT!*` to tell if it is going round in circles. The default limit `SIMPLIMIT!*` on `SIMPCOUNT!*` is usually either 1000 or 2000 (it is set at two places in the source, in ALG1, with different values, of which the second, 1000, is the one currently used in the Atari version) and if that is exceeded one gets the well-known message "`Simplification recursion too deep`".

The arguments to `REVAL1`, `SIMP!*` and `SIMP` can be either pseudo-prefix or prefix forms, and already simplified standard quotients (in pseudo-prefix form and marked by the final `T`) are not re-simplified. It

---

[10] In some implementations, the `MODULE` and `ENDMODULE` commands which subdivide the REDUCE source as described at the beginning of this section also have a mode-changing side-effect which can be confusing.

should be noted that if one wishes to add simplification functions to the standard simplifier these should return a standard quotient, and the arguments they will receive depend on their flags (see below).

What follows is an attempt to briefly describe the main features of the functions called by SIMP.

1. There is a series of special simplifiers called **SIMPxxx**, where **xxx** is one of: **ABS**, **ATOM**, **CAR**, **DIFF**(erence), **EXPT**, **IDEN**(tifier), **MINUS**, **PLUS**, **QUOT**(ient), **RAD**(ical), **RECIP**(rocal), **SET**, **!*SQ**, **SQRT**, **SUB**, **TIMES**, **TRIG**. All of these may recursively call **SIMP** and many are found as the **SIMPFN** properties of the corresponding operators, which **SIMP** looks for if it encounters a name which might be an operator. (Note that if the **SIMPFN** is flagged **FULL** the operator name is passed as its first argument: otherwise, except for **SIMPIDEN**, it gets only the arguments of the operator.)

2. There are functions for arithmetic on the various parts and levels of standard quotients. Some of the principal ones are as follows:

   for SQ expressions: ADDSQ, CANONSQ, EXPTSQ (and EXPSQ), INVSQ, MULTSQ, NEGSQ, QUOTSQ, and so on. One special function is CANCEL which calls COMFAC (for common factors) and GCDF etc.;

   for standard forms: ADDF, EXPTF and EXPF, MINUSF, MULTF, QUOTF, NEGF, ...;

   for powers divided by standard forms: MULTPQ, MULTPF;

   for domain elements divided by forms: ADDD, MULTD, QUOTD, QUOTDD;

   for domain elements: ADDDM, MULTDM;

   for kernels: QUOTK, GCDK.

   A special function **SUBF** implements the **SUB** command.

   Most of these call related commands of the form *name*!* which check the arguments are of the correct type and then pass them to functions called *name*1 which do the real work.

3. Substitutions are checked for by a function **SUBS2** which calls a considerable list of other functions, notably **SUBS2Q**, **SUBS3Q**. For instance, using $\cos^2 x \to 1 - \sin^2 x$ in $\cos^4 x + \cos^2 x \sin^2 x$ calls **SUBS2Q** 9 times, and the Fourier analysis in the exercises calls **SUBS3Q** 97 times. The matching functions for substitutions work on prefix forms.

Reordering of variables and kernels is done by a function REORDER according to the order set by KORDER (which is held on a list KORD!*) and this can call subsidiary functions REPLUS, RETIMES, RADDF, RADDSQ, RMULTPF ....

4. There are building functions for the final output form: MKSQ, MK!*SQ, PREPSQ, PREPF, MKSP, MKSP!*, MKSFPF ...
Note that conversions from prefix form to standard quotient form and vice versa are used internally in simplification so that each function receives its arguments and returns its value in the correct format.

5. Operators may have OPMTCH called which looks at the OPMTCH property of the operator and tries to match the arguments in order to evaluate the expression.

6. Special functions for matrix, vector and operator expressions may be called (e.g. the functions implementing LINEAR).

6.12 Substitutions

CLEAR, LET, MATCH, SAVEAS and SETK (i.e. ":=" in algebraic mode) all work via internal functions LET2 and LETEXPRN. The difference between an unconditional LET and SETK when the left side is an identifier is that LET does not evaluate its right side: both set the AVALUE of the variable. Otherwise LET2 and LETEXPRN maintain a number of association lists (a-lists) which are checked by the internal substitution functions called by the simplification routines (see above). One of these is !*MATCH whose entries have the form

(U . (T *or* NIL . *mcond or* T . V . NIL))

if U is to be substituted by V. !*MATCH is the list used for all substitutions which need to match more than one simple variable or a power. Here the first entry in the value part of the pair is a flag showing if a partial match is allowed (i.e. whether LET or MATCH was called), *mcond* is the condition (if any), and the final NIL is the initial value of a list later used to store those parts of the original U which are already matched. Similar lists are held as the OPMTCH property of an operator, or on POWLIS!*, POWLIS1!* and ASYMPLIS!* for substitution of powers (POWLIS!* for simple substitutions, POWLIS1!* for conditional substitutions, and ASYMPLIS!* for substitutions giving zero).

Note that by default (cf. §2.8) let-rules such as LET a/b = x are translated as LET a = b*x, and LET x = y where x is a polynomial becomes LET *leading term of* x = -*rest of* x + y.

Let-rules set by a FOR ALL command have results similar to those from simple let-rules except that each identifier to which the FOR ALL applies is replaced by a similar identifier with the prefix = (which must be escaped by !, so that x becomes !=x).

6.13 Exercises

1. Evaluate, if possible:
(a) CAR('((A B) (C D)));
(b) CDR('CDR, (CAR('(A B))));
(c) CADR('(A (B C) D));
(d) SET('TOOLS, CONS('HAMMER, 'SCREWDRIVER));
[If you do these by running REDUCE, make sure you understand why the result is obtained.]

2. Explain the action and purpose of the following functions:

```
symbolic procedure function1 s;
   if null s then 0
   else if atom(s) then 1
   else function1(car s) + function1(cdr s);

symbolic procedure function2(u, v);
   if null u then v
   else if atom u then function2(list u, v)
   else if atom v then function2(u, list v)
   else if member(car u, v) then function2(cdr u, v)
   else car u . function2(cdr u, v);
```

3. Write RLISP procedures to perform the following tasks:

(a) form the intersection of two lists of elements,

(b) given a list containing lists as elements, make a single list of all atoms appearing in any sub-list at any level (atoms may be repeated),

(c) delete repeats from a list, and

(d) combine (b) and (c).

4. The following definition is part of the REDUCE 3.3 code to implement the PREPSQ and PREPF functions. (The code for EXCHK is in ALG2 and this is an exact copy, slightly reformatted to fit on the page.) At top level it is called with the following arguments: U is a standard term, V and W are NIL. You may assume that the action of the functions ASSOC2, ADDSQ, EQCAR, PREPSQX, SIMPEXPON and SQCHK is as follows:

ASSOC2 is like ASSOC but checks the second element of each pair on an a-list as the key.

ADDSQ adds two standard quotients and returns a standard quotient.

EQCAR(u,v) is essentially CAR u EQ v (see the Appendix, §E).

PREPSQX acts on a standard quotient and returns the equivalent prefix form.

SIMPEXPON simplifies exponentials.

SQCHK checks if a quantity is a standard quotient and returns the prefix form.

Describe the purpose of EXCHK, how it works, and what it returns.

```
symbolic procedure exchk u; exchk1(u,nil,nil,nil);

symbolic procedure exchk1(u,v,w,x);
   % checks forms for kernels in EXPT.
   % U is list of powers.
   % V is used to build up the final answer. W is an
   % association list of previous non-constant (non
   % foldable) EXPT's, X is an association list of
   % constant (foldable) EXPT arguments.
   if null u then exchk2(append(x,w),v)
      else if eqcar(caar u,'expt)
         then
            begin scalar y,z;
               y := simpexpon list('times,cdar u,
                                   caddar car u);
               if numberp cadaar u  % constant arg
                  then <<z := assoc2(y,x);
                     if z
                        then rplaca(z,car z*cadaar u)
                       else x := (cadaar u . y) . x>>
                  else <<z := assoc(cadaar u,w);
                     if z
                       then rplacd(z,addsq(y,cdr z))
                       else
                          w := (cadaar u . y) . w>>;
              return exchk1(cdr u,v,w,x)
           end
         else if cdar u=1
           then exchk1(cdr u,sqchk caar u . v,w,x)
           else exchk1(cdr u,
                list('expt,sqchk caar u,cdar u)
                . v,w,x);
```

```
symbolic procedure exchk2(u,v);
   if null u then v
      else exchk2(cdr u,
         ((if eqcar(x,'quotient) and caddr x = 2
               then if cadr x = 1
                  then list('sqrt,caar u)
                  else
                     list('expt,
                          list('sqrt,caar u),cadr x)
               else if x=0.5 then list('sqrt,caar u)
               else if x=1 then caar u
               else list('expt,caar u,x))
                 where x = prepsqx cdar u)
          . v);
```

5. In REDUCE, the function SUBS3F1 calls the function SUBS3T on the leading term of a standard form. This in turn calls the function MTCHK, usually with arguments U, V where U is the CAR of the standard term, and V is a list, and then may call SUBS3F1 recursively on the CDR of the term, with a new list V returned by MTCHK. At the first call of MTCHK, V is the list !*MATCH (usually).

   By executing the following sequence of REDUCE commands and inspecting the results, deduce what MTCHK does. (You may also look at the source code, which is near the start of ALG2.) [Remember that TRACE or TR may need to be replaced by a related Lisp command appropriate to your implementation – see §6.9: the version using TRACE works for PSL and Cambridge REDUCE. Beware also that the output is long – the first few parts are enough to solve the problem.]

```
TRACE '(SUBS3T MTCHK);  % or TR SUBS3T, MTCHK; or ...
% The rest is from the standard REDUCE test file
FOR ALL X,Y LET
    COS(X)*COS(Y) = (COS(X+Y)+COS(X-Y))/2,
    COS(X)*SIN(Y) = (SIN(X+Y)-SIN(X-Y))/2,
    SIN(X)*SIN(Y) = (COS(X-Y)-COS(X+Y))/2,
    COS(X)**2 = (1+COS(2*X))/2,
    SIN(X)**2 = (1-COS(2*X))/2;

FACTOR COS,SIN;

(A1*COS(WT) + A3*COS(3*WT)
     + B1*SIN(WT) + B3*SIN(3*WT))**3;
```

7

Factorization and integration in REDUCE

The purpose of this chapter is to give an introduction to the algorithms employed by two of the most important of REDUCE's algebraic packages, namely the factorizer and integrator, without, however, going in full detail into the underlying pure mathematics. We do not give details of the bigfloat package, although it is one of the largest pieces of REDUCE source code, nor of the several smaller packages such as the internal editor, matrix handler and so on.

7.1 Factorization

To understand the method used by REDUCE for factorization we need to understand the finite fields $GF(p)$ mentioned but not defined at the start of chapter 5. They arise as interesting examples of a general construction which starts with a map f of a ring R into a ring S that preserves all the ring operations. The map f is then called a *ring homomorphism.* The set I of all elements r in R such that $f(r) = 0$ in S (called the *kernel* of the map f) is itself a ring, a subring of R with special properties, namely that if a and b are any elements of I and r is any element of R, then $a + b \in I$ and $a * r \in I$: a subring I in R with such properties is called an *ideal* in R. The subset of all elements of S which are the image $f(r)$ for some r in R also forms a ring (a subring of S), called the *quotient ring* and denoted by R/I.

We have already encountered an example of this construction in the construction of an algebraic extension field $E[x]/(p(x))$. A well-known particular case is when E is the real numbers $\mathbb{R}$, and p is the polynomial $x^2 + 1$, for which the quotient ring is the field $\mathbb{C}$ of complex numbers.

$GF(p)$ is the case where R is $\mathbb{Z}$ and I is the set, denoted (p), of all integers divisible by the prime p. Note that although $\mathbb{Z}$ is not a field, $\mathbb{Z}/(p)$ is, as is easily checked. REDUCE's modular arithmetic, described in chapter 1, will of course become arithmetic in $GF(p)$ when the modulus is a prime p. The factorizer uses this facility internally, and can factorize over $GF(p)$ and algebraic numbers as well as over the integers (except that it does not at present do multivariate factorization over $GF(p)$).

The importance of $GF(p)$ for factorization (where by factorization we mean that we seek factors with coefficients which are integers, which can be shown to be equivalent to factorization over the rationals), is that one can do it by first factorizing an expression over $GF(p)$ and then translating the result back into a factorization over $\mathbb{Z}$. The first of these steps (in REDUCE) uses a method due to Berlekamp (large parts of which had earlier been described by Schwarz (1956)) and the second uses extensions of a method due to Hensel (for references see Davenport *et al.* 1988, Buchberger *et al.* 1983). The similar method for multivariate GCDs is called the EZGCD method (EZ standing for "extended Zassenhaus", since it uses the extension of Hensel's lifting method due to Zassenhaus). It was introduced for multivariate GCDs by Moses and Yun (1973) and for factorization by Wang and Rothschild; REDUCE now actually uses Wang's revised version of this, introduced in Wang (1978). The factorizer uses the EZGCD method for the GCDs it needs, and this can lead to some confusion since the trace messages from the Hensel steps refer to factorization even when a GCD calculation is actually being performed.

Berlekamp's method only works when there are no repeated factors (in which case the polynomial is called "square-free"), so the method first preprocesses the input polynomial, the resulting expression in terms of square-free factors being called the square-free factorization, a term which can seem confusing since such a factorization may involve squares and higher powers of square-free factors. (It should be noted that there are other, possibly more efficient, methods than Berlekamp's for passing from a square-free factorization to a complete factorization over $GF(p)$; these are not implemented in REDUCE.)

There is only room for a sketch of the Berlekamp and Hensel methods here; a more detailed account is given in Davenport *et al.* (1988) but the actual methods used by REDUCE differ in some respects from those described there and in particular Davenport *et al.* do not give a full description for the multivariate case, in which REDUCE follows the methods described in Wang (1978).

Suppose we are attempting to factorize a univariate polynomial $q(x)$. The first step is to take out the GCD of the coefficients of the powers of x, which is called the *content* with respect to x, as mentioned in chapter 2. (In the case of a multivariate polynomial the content with respect to x will be, in general, a polynomial in the other variables, which would then have to be factorized by recursive application of the factorizer.) For a univariate polynomial the content will be a rational number. The other factor, now with content 1, is called the *primitive part.*

To reduce q to square-free factors one first divides q by the greatest common divisor (GCD) of q and q'. The result, C_1, is the product of each square-free factor taken just once, while the GCD obtained, D_1, is the product of any repeated factors taken to one lower power than in q. One can use an obvious recursion of this step, or a more complicated method due to Yun (described in Kaltofen 1983) also using derivatives, but REDUCE actually avoids further differentiation by using the following method due to Musser. Repeatedly take $D_{i+1} = \gcd(C_i, D_i)$, $C_{i+1} = C_i/D_{i+1}$, $P_i = D_i/D_{i+1}$ until $D_i = 1$. Then P_i is the product of all the square-free factors which appear to exactly the ith power.

Now we may suppose $q(x)$ is square-free. We can compute an upper bound on the size of the coefficients in any of the factors, using the results of Landau and Mignotte; although this is usually too conservative, it is the best possible. Then the method is essentially to choose a prime number p not dividing the leading coefficient of $q(x)$ and such that $q(x)$ is square-free mod p; factorize $q(x)$ mod p by Berlekamp's method; from this find a factorization mod p^k by Hensel's method where k is chosen such that p^k is greater than twice the Landau-Mignotte bound; and finally test the resulting factors (which are correct mod p^k and will have correct coefficients if they are genuine factors) first singly, then pairwise, then in groups of 3, and so on until all of them have been used in some genuine factorization of $q(x)$ over $\mathbb{Z}$. A small subtlety arises from the need to first check that the factors mod p^k have been adjusted so that their overall numerical factor is correct. This is done by adjusting them to be monic and then multiplying both q and the supposed factor by the leading coefficient of q before the trial division, taking the primitive part of any resulting factor as the true factor.

The Berlekamp method depends on the fact that if $q(x)$ is a square-free primitive polynomial (mod p) with irreducible factors q_k for $1 \leq k \leq r$ then the ring $K[x]/(q(x))$ (where K in our case will be $GF(p)$) is homomorphic to a vector space which is a direct sum of the rings $K[x]/(q_k(x))$ and each of these summands is a field (this is a way of stating the "Chinese remainder theorem"). Now the Frobenius map $x \rightarrow x^p$ in such fields leaves only the constants, i.e. elements of $GF(p)$, fixed. A polynomial v whose image in a particular $K[x]/(q_k(x))$ is zero must be divisible by that $q_k(x)$ and if its images in the similar fields for other k are non-zero constants then $q_k = \gcd(q, v)$. The strategy is thus to find non-trivial v in $K[x]/(q(x))$ which are fixed under the Frobenius map, and try subtracting the possible constants and finding all non-trivial GCDs. Frequently one such v gives all the irreducible factors, but if not then any non-trivial GCD found this way will be a product of irreducible factors and can be (and in REDUCE is) used to break the problem down into simpler problems.

Thus the method handles the rings involved as vector spaces and begins by computing a matrix Q whose k^{th} row is given by the coefficients in the pk^{th} power of x modulo ($q(x)$ and p), the kth row therefore representing the image in $K[x]/(q(x))$ of x^k under the Frobenius map. The eigenvectors v of this matrix with eigenvalue 1 are found: the number of these is the number of factors [because it is the number of copies of $GF(p)$, one in each $K[x]/(q_k(x))$]. REDUCE takes 5 randomly chosen small primes and finds this number of factors for each prime (the number 5 was chosen on the basis of experiments by D. Musser). From these it chooses 3 which give the lowest number(s) of factors and completes the factorization mod p for each. Then REDUCE checks that these factorizations are compatible and if so continues using the first of them. The check uses the so-called degree sets which are the leading powers of the factors found.

In fact, REDUCE finds the linear factors (mod p) before starting this process, so the Berlekamp method is only used for the nonlinear factors. The linear factors mod p are found by taking

$$\gcd(q, x^p - x),$$

which gives their product, as can be seen from the properties of the Frobenius map, and then trying different constants until all the zeros of this polynomial have been found. (This is perhaps the point at which to note that "cyclotomic" polynomials, those of the form $x^n \pm 1$, also get special handling.)

The Hensel "lifting" can be done in a way that doubles the exponent of p at each step. Suppose $q(x) = g_1(x)h_1(x) \bmod Q$ (where Q is some power of p) and is $g_2(x)h_2(x) \bmod Q^2$, where $g_2 = g_1 + Q\bar{g}_2$ for some $\bar{g}_2$ and similarly for h and q. Then we have the equation

$$\frac{q_1 - g_1h_1}{Q} + \bar{q}_2 = \bar{g}_2h_1 + \bar{h}_2g_1 \pmod{Q}$$

where q_1 is the image of q mod Q. Since g_1 and h_1 are relatively prime (as the polynomial q at this stage is square-free) we can use the extended Euclidean algorithm[1] to find the corrections $\bar{g}_2$ and $\bar{h}_2$.

---

[1]Readers unfamiliar with this and Bezout's identity will find the details in many texts, e.g. in the appendix to Davenport *et al.* (1988).

(There are some subtleties which allow one to go by a similar step from the coefficients $a(x)$ and $b(x)$ in Bezout's identity $a(x)g(x)+b(x)h(x) = 1 \bmod Q$ to the corrections required to give the similar coefficients mod Q^2, and hence we need to perform the Euclidean algorithm only once, namely mod p.) The same ideas apply to cases with more than two factors, the displayed equation above being modified so that if the factors are denoted by g_k $(0 \leq k \leq r)$ then the correction to g_i is multiplied by the product of all the other factors g_j $(j \neq i)$.

Here are some excerpts from a simple factorization (of $x^7 - x^6 + x^2 - 1$). The messages are from the output with the `TRFAC` and `OVERVIEW` switches on, except where otherwise noted. The system begins by establishing the variables and the basic nature of the polynomial.

```
The polynomial is univariate, primitive and square-free
so we can treat it slightly more specifically. We
factorise mod several primes,then pick the best one
to use in the Hensel construction.
Degree sets are:
       2 4 1
       3 3 1
       3 3 1
Possible degrees for factors are:
1 3 4 6
The chosen prime is 23
The polynomial mod 23, made monic, is:
 7       6    2
x  + 22*x  + x  + 22
and the factors of this modular polynomial are:
 3       2
x  + 12*x  + 22*x + 14
 3       2
x  + 11*x  + 7*x + 5
x + 22
```

Now one must apply the Hensel lifting. For the example, REDUCE reports:

```
We are now ready to use the Hensel construction to grow
in powers of 23
Initial factors mod 23 with some correct coefficients:
       3       2
 f(1)=x  - 11*x  - x - 9
       3       2
 f(2)=x  + 11*x  + 7*x + 5
 f(3)=x - 1
Univariate factors, possibly with adjusted leading
coefficients, are:
 f(1)=x - 1
       6
 f(2)=x  + x + 1
The univariate factors are:
```

```
x - 1
 6
x  + x + 1
```

The corrections of the modular results essentially have to do with, for example, the fact that −45 is equivalent to 1 mod 23 but clearly the factorization over $\mathbb{Z}$ cannot involve 45 as a constant term in any of the factors.

The coefficient bound is shown in the output from `TRALLFAC` (the switch for more detailed tracing) and is

```
Coefficient bound = 110
```

Even fuller details of the calculations can be obtained by turning on the switch `TIMINGS`.

For the case of multinomials, the idea of a factorization mod p is extended by the idea of evaluations $x_i = a_i$ in which all variables (except one) are replaced by a numerical value (the set of these values is called an image set). The image polynomial obtained by these evaluations is factorized as a univariate polynomial. The image set is chosen, with an associated prime for use in the univariate factorization, so that the leading coefficient does not have a zero image, the image polynomial is square-free, and all factors of the leading coefficient have images such that there are primes dividing exactly one of these images. (The factorization of the leading coefficient is of course done by a recursive call of the factorizer.) This way of choosing an image set allows the factors of the leading coefficient to be correctly distributed between the full factors. A Hensel lifting has to be defined which enables one to put back the x_i replaced by the a_i and this is done (see Wang 1978) by recursively putting back each variable in turn, and by doing it in successive powers of that variable. The method also attempts to use any available deductions about other true coefficients and tries bivariate factorizations in different pairs of variables. It involves a highly recursive program of too great a complexity to be fully described here.

7.2 Integration

Since the time of Liouville it has been known what form an integral of a given integrand would take if it were possible to find such an integral in terms of elementary functions. As with the discussion of representations and factorization, it helps to give a few parts of the formal framework (Kolchin 1973, Risch 1969, Davenport 1981).

Definition: A *differential field* is a field (in the sense of algebra) equipped with a differentiation operation, say $'$, obeying the usual rules for differentiation of products and sums.

A differential field may have more than one differentiation (if it represents functions of more than one variable), but we need only consider one here. Note that the chain rule is *not* assumed, since we have made no requirement that the field represent functions and the concept of a function of a function thus has no meaning at this stage. Every differential field F contains a sub-field of constants consisting of those elements f of F which obey $f' = 0$. We shall denote this field (which, since it is easily shown to contain 1, is not trivial) by K and assume that it is one of $\mathbb{Q}$, $\mathbb{R}$ or $\mathbb{C}$ as appropriate.

Definition: An *elementary extension* of a differential field F is a differential field $F(\theta)$ which is an extension of F such that θ is either
1. algebraic over F,
2. a logarithm over F, or
3. an exponential over F,

where θ is a logarithm if $\theta' = f'/f$ for some $f \in F$ or an exponential if $\theta' = f\theta$ for some $f \in F$.

If θ is transcendental over F and either an exponential or a logarithm, then $F(\theta)$ is called a *transcendental elementary extension* of F. We now have a precise definition of the elementary functions: they are exactly those functions which are members of some field obtained by a tower of elementary extensions of the rational function field $K(x)$, where in practice we are usually referring to $K = \mathbb{Q}$ or, perhaps after algebraic extensions, to $K = \mathbb{C}$ or one of its subfields. This includes the exponential and logarithmic functions (obviously), and the trigonometric and hyperbolic functions.

Liouville's result says essentially that if an integrand f is in a certain differential field F, and has an elementary integral over that field, the integral will be of the form

$$v_0 + \sum_{i=1}^{n} c_i \log v_i$$

where v_0 is in F, the v_i are in some extension of F by constants algebraic over the initial K, and the c_i are in this extended constant field; the v_i are essentially factors of the denominator of f. The problem remains of how to actually find the c_i and v_j. For rational functions, i.e. elements of $K(x)$, Hermite, in a method later improved by Horowitz, showed how to solve this, and the handling of the logarithmic part in that method was more recently improved by Rothstein and Trager (details of these methods are given in Davenport *et al.* (1988)). Risch (1969) gave an algorithm for the case of transcendental elementary extensions of the rational functions; this algorithm is recursive in the extensions, and quite complicated. Before applying it one has (in principle) to check that the different extension variables are not algebraically related, but fortunately there are so-called "structure theorems" which provide methods to check this.

For rational functions Risch's method is the same as Horowitz's (plus Rothstein's). For an extension $E(\theta)$ in the tower one can write the integrand in a partial fraction form, and do the same, with unknown coefficients, for the possible integral (using Liouville's result to specify the form). Then by differentiating and equating coefficients one can, when θ is a logarithm, reduce the problem to integration in E (which can recursively either be performed or proved impossible in elementary terms) together with an application of Rothstein's method in $E(\theta)$. The case where θ is an exponential follows basically the same ideas, but there are some subtle differences in the definitions of the integrals to be done in E and one new feature, namely that some of the coefficients to be found are given by a first-order linear differential equation in E, the "Risch differential equation". This equation would be trivial to solve by the usual integrating factor method were it not for the fact that such a method immediately leads back to the very problem (integration in $E(\theta)$) that we were trying to solve initially, and one needs a more subtle approach to avoid that. Details can be found in Davenport *et al.* (1988) but are omitted here because REDUCE does not use this method.

The problem of algebraic functions turned out to be much more difficult. Davenport (1981) gave a solution, but this has only been implemented for the case of square roots. It is that program which is distributed with REDUCE as the ALGINT package[2]. (To use it, one usually has to load it, and it is then invoked automatically whenever `INT` is called. However, the extra facility can be disabled by turning off the `ALGINT` switch.) Trager has given an alternative to Davenport's method. Recently Bronstein (1990; see also Bronstein *et al.* 1989) has announced a method for handling the still more difficult case of mixed algebraic and transcendental functions but this has not yet been implemented in REDUCE. (Dr. Jenks' lecture at the school discussed the Scratchpad II implementation.)

The method used by the command `INT` in REDUCE (when `ALGINT` is not in use) is not in fact the (original) Risch method, but is instead a different method developed by Risch and Norman, which was

---

[2]See §§8.1, 8.2 for further details and an example.

intended to be more efficient and simpler to program. Unfortunately it turned out that the Risch-Norman method is incomplete (unlike the first Risch method) in that it may fail to find an elementary integral when one exists, and so its failure is no proof of non-integrability. (It is not entirely clear whether the Risch-Norman method can be made complete.) However, it has, after some amendments, proved quite efficient and reliable, and can integrate some functions not strictly covered by the Risch method.

If one uses the integrator `INT` with the trace switch `TRINT` on, one will see a series of messages which explain exactly what is happening. These correspond to the steps of the procedure which can be described as follows. (The REDUCE extracts here are taken from the integration of

$$\frac{(2/x)\log x + 1}{1 + (\log x)^2} - \frac{2\log x}{(1 + (\log x)^2)^2}$$

and show only a part of the messages given.)

1. The first step is to scan the integrand and work out a suitable field from which to begin; this is a matter of choosing the "field descriptors" z_i, i.e. a set of quantities such that the integrand is in a suitable member of a tower of extensions $K(z_0, z_1, \ldots, z_n)$ of a constant field K, so that each z_j extends the field $K(z_0, \ldots, z_{j-1})$. Naturally the extension variable z_0 is taken to be the variable of integration. Moreover, the order of the z_i (shown by their numbering) is significant as the method uses an ordering of terms based on this, taking the highest-numbered z_j as the leading one. For the example that we are considering, the field descriptors are x and $\log x$, with $\log x$ as the leading one, and REDUCE says

```
extension variables z<i> are
((log x) x)
```

(note that the z_i appear with the leading one first rather than last). The integrand is put over a common denominator at this stage, which will put an x into the denominator of our example, and REDUCE also "unnormalizes" the integrand by calculating the derivatives of the field descriptors and adding extra factors to the numerator and denominator so that all factors needed in denominators to match the derivatives of the field descriptors will be present (this is important in solving certain integration problems, such as example 8 of Geddes and Stefanus (1989)[3]).

2. Each irreducible factor v_j of the denominator might in principle have an associated logarithmic term in Liouville's form of the integral, if that integral actually exists. One could therefore try to factorize the denominator into irreducible factors, if necessary adding algebraic constants to do so. Actually this often goes too far, because there may be no related logarithmic term (as an example, consider $1/x^2$) or because the terms that arise combine in a form that enables one to avoid the algebraic extensions (as an example, consider $x/(x^2+1)$, where extension by i can be avoided). One could find the really necessary set of terms by Rothstein's method or some variant of it; in practice, integrands often factorize readily as far as is required. `INT` has its own factorization procedures for this purpose, distinct from the main factorizer; if these fail, a message, telling the user that

   ```
   The following has not been split
   ```

   followed by the relevant polynomial, is generated.

   `INT` actually deals with quadratic factors which have complex roots without introducing i; instead it uses a linear combination of the logarithm of the quadratic and the inverse tangent function, `ATAN` in REDUCE, which is a combination of complex logarithms. For the example the v_j $(j \geq 1)$ terms are thus

   $$c_1 \log((\log x)^2 + 1) + c_2 \tan^{-1} \log x + c_3 \log x$$

   and REDUCE gives a list of these in a prefix form, with the arguments of the `LOG` and `ATAN` functions given as standard quotients and the constants, written after the function name, as standard forms:

   ```
   loglist
   ((log (((c1 . 1) . 1)) ((((log x) . 2) . 1) . 1) . 1)
    (atan (((c2 . 1) . 1)) ((((log x) . 1) . 1)) . 1)
    (log (((c3 . 1) . 1)) (((x . 1) . 1)) . 1))
   ```

[3]It is worth noting that by allowing a wider range of field descriptors, including sine and cosine (Fitch 1981), the other cases which Geddes and Stefanus describe as unsolved by the Risch-Norman method are not only solved by REDUCE's `INT` but given in a nicer form than Geddes and Stefanus give.

The numbering of the constants c_i is not the same in all implementations.

3. At the same stage one can find the denominator of the v_0 part of Liouville's form. (REDUCE actually gives this result before writing the logarithm part.) It will be a product of the factors of the denominators of the integrand with each taken to one less power, except for any factor which is an exponential: these latter are taken to the same power. In the example the denominator of v_0 will be

$$(\log x)^2 + 1$$

and REDUCE says

```
denominator of 1st part of answer is:
factors of multiplicity 1
      2
log(x)  + 1
trivial factor found
(x)
```

(The x is trivial because it is a field descriptor, not an exponential, and appears in the denominator of the integrand only to the first power.)

4. The numerator of the v_0 part is now taken as a multinomial u in the field descriptors, with coefficients named $u_{ijk\ldots}$. Thus for the example we will have

$$v_0 = \frac{\sum u_{kj}(\log x)^k x^j}{(\log x)^2 + 1} .$$

Combining this with the expression from step 2 above gives the whole of the integral (if one exists): the coefficients c_j and $u_{ijk\ldots}$ remain to be determined. This is done by differentiating the supposed integral, putting the result over a common denominator, which should match the denominator of the (unnormalized) integrand, and (in the next step) equating coefficients of powers of the field descriptors in the numerators. The parts of this equation involving u (i.e. v_0) are put on one side and the rest (the integrand less the derivatives of the logarithms) on the other, this latter being called the residue. The $u_{ijk\ldots}$ part can be written in two ways, the first by collecting all terms with the same u indices $ijk\ldots$ together and the second by collecting like powers of the field descriptors together. REDUCE reports these steps by giving the derivative of the logarithm terms multiplied by the overall denominator of the integrand (in "unnormalized" form), the factor multiplying a coefficient $u_{ijk...}$ and the residue, as polynomials in the field descriptors. Here are these three expressions for the example:

```
************ 'derivative' of logs is:
       4               3               2
log(x) *c3 + 2*log(x) *c1 + log(x) *c2 +

              2
2*log(x) *c3 + 2*log(x)*c1 + c2 + c3

distributed form of u is:
       j         k + 2                     j             k + 1
 j*(x  *log(x)        ) + ( k - 2)*(x  *log(x)               ) +

       j          k           j           k - 1
 j*(x  *log(x)  ) +  k*(x  *log(x)              )
distributed form of l.h.s. is:
               4                                  3
- c3*(log(x)  ) + ( - 2*c1 + 2)*(log(x)  ) +

           2                                  2
 (log(x)  *x) + ( - c2 - 2*c3)*(log(x)  ) -

2*(log(x)*x) + ( - 2*c1 + 2)*(log(x)) + (x) - c2 - c3
```

5. Now one equates coefficients of terms to solve for the $u_{ijk...}$ and the c_i successively, starting from the highest terms in the residue (using the chosen ordering of field descriptors). Here we come to the most awkward point in the procedure: so far we have not put any bound on the order of the terms in the $u_{ijk...}$ sum, so that in principle we have an infinite number of equations. A bound must therefore now be chosen, and unfortunately although any elementary integral will only involve a finite number of terms no theorem giving a universally applicable bound on this number is known. The rule currently adopted by INT is the "superbound" introduced by Fitch (1981).

   The method for the superbound is the following. For each field descriptor z_i, 1 is added to the highest power of z_i in the residue, giving m say, and the distributed form of the terms from u given in step 4 is then inspected for terms coming from $(z_i)^j$ with a coefficient factor of the form $(j-n)$ for some integer n. If $n > m$, n is used instead of m. The highest terms allowed are those with all field descriptors taken to their corresponding superbound powers, and all lower terms are considered. (There was a more complicated but smaller set of allowed terms in Norman's ISIS system, but we have been unable to find documentation of the procedure used there.)

In the example, the initial residue has terms involving $(\log x)^4$ and x so the initial estimates of the bounds would be 5 and 2 respectively. The only $(j-n)$-type factor is the $(k-2)$ shown in step 4, and since $5 > 2$, 5 remains the bound. REDUCE reports:

```
maximum order determined as
(5 2)
```

Now REDUCE tries to solve the equation obtained by taking all terms arising from u which match the highest powers in the residue. If this gives the leading $u_{ijk...}$ uniquely, REDUCE solves for it. If not, INT creates new c_i names for as yet unknown $u_{ijk...}$ of lower order, and writes the higher ones in terms of them, as it does in our example (where the first equation to be solved is for the $(\log x)^4$ terms and involves u_{50} and u_{30}). The new c_j terms are transferred to the residue, and removed from the expression arising from u. In our example, REDUCE reports:

```
***** u(3 0) =
c4
introduced residue ((4 0) (((c4 . 1) . 1)) . 1)
introduced residue ((2 0) (((c4 . 1) . 3)) . 1)
constant map changed to ((c4 . 4) (c3 . 3) (c2 . 2)
 (c1 . 1))
***** u(5 0) =

    c3 + c4
 - ---------
      5
```

6. When a new term in v_0 has been fixed, all the terms associated with that $u_{ijk...}$ are transferred to the residue. (In our example, at the first step, these will be all terms coming from u_{50}.) Obviously if the residue is then zero, the integration is ended. If terms higher than the previous bound appear in the residue, their coefficients are equated to zero (INT calls such an equation a "c-equation" because it will not involve any $u_{ijk...}$, except of course, indirectly, those for which extra c_j were introduced). If this fails, there is (according to REDUCE) no integral (the incompleteness of the method lies precisely in the fact that this inference may be wrong,

or at least has not yet been proved to be correct). Otherwise one can iterate step 5, recalculating the superbound with the new residue and using the new set of terms from u (i.e. ignoring in u those terms transferred to the residue). In our example, this iteration goes through quite a few cycles, details of which would simply occupy space without illustrating new principles, but eventually succeeds.

This integrator is far from complete for transcendental functions (see Fitch 1981) but works well on the whole. There are experiments in progress which may improve its efficiency noticeably, e.g. it may pay *not* to take a single common denominator, but to separate terms dependent on different descriptors in the integrand and integrate each part separately (a "distributed" version): the improvement would come from the reduction in dimension of the search space of the $u_{ijk\ldots}$. However, progress that would make it universally correct seems unlikely, and in the long term it might be better to add a full version of the original Risch algorithm extended to algebraic and mixed functions, possibly retaining the Risch-Norman algorithm as a first method for efficiency reasons, as has been proposed for Maple (Geddes and Stefanus 1989).

7.3 Exercises

1. (This question was suggested by Peter Kropholler.)
   The symbolic determinant

$$determ = \begin{vmatrix} 2a & b & c & d \\ b & 2a^p & b^p & c^p \\ c & b^p & 2a^{p^2} & b^{p^2} \\ d & c^p & b^{p^2} & 2a^{p^3} \end{vmatrix}$$

   where $p = 3$ and calculations are to be done over GF(3), is obviously quadratic in d. Show that the discriminant of that quadratic is a perfect square (in fact it is a perfect quartic), construct its square root, and hence find the values of d for which *determ* vanishes, as rational functions. [Note that since multivariate factorization using modular arithmetic is not provided, one must find another method for the decomposition of the discriminant into factors. One must also be careful about when `MODULAR` should be on and when it should be off (e.g. one does not want 4 to be considered as 1 mod 3 when using it as an index to set a matrix element).]

2. With `TRFAC` on, interpret the output of factorizing the following functions:
   (a) $x^4 + 1$
   (b) $x^4 + 3x^3 - 13x^2 + 6x - 30$
   (c) $x^8 + x + 1$
   (d) $2x^4 + (y^2 + 10z^2)x^2 - 6y^4 - z^2y^2 + 12z^4$
   (e) the expansion of $(z + xy + 10)(xz + y + 20)(yz + x + 20)$.

3. With `TRINT` on, use REDUCE's `INT` on the following functions and explain the results:
   (a) $\exp(x^2)$
   (b) $\log x/x^2$
   (c) $xe^x \log x$
   (d) $\exp(1 + (\log(z^3 - 1))/2)$.

8

The user-contributed REDUCE packages

The capabilities of the basic REDUCE system are being progressively extended by a library of "packages", written mostly in symbolic mode by "users" rather than those primarily developing REDUCE. These packages currently fall into two classes: those that are distributed with the system (since version 3.3), and those that are not distributed but are available to REDUCE users. The former have mostly existed for longer and been more widely tested than the latter, and have a slightly more official status. The latter have a similar status to any other public domain software, and are available via electronic mail from the REDUCE network library (Internet address: `reduce-netlib@rand.org`)[1]. This chapter provides a brief sketch of some of the packages, and a more detailed introduction to the Gröbner Basis package for manipulating coupled polynomial systems.

Both this chapter and much of the next apply *in detail* only to REDUCE 3.3 (and later versions, to some extent) due to internal changes between REDUCE versions. Programs written for REDUCE 3.3 may need modification to run under another version of REDUCE and vice versa. However, we expect the currently distributed packages to be maintained under future versions of REDUCE; they may also be developed further and additional packages may be distributed.

8.1 The distributed user-contributed packages

Six packages are currently supplied with the standard REDUCE 3.3 distribution. The precise form in which the packages are available at any particular site is very system dependent, but at the very least the standard packages should be available as source code together with

[1]See the Appendix, §F, for further details.

some online documentation and test or demonstration files. There appears to be no guidance provided on exactly what to do with the source files, so we will give some below.

These are the standard packages:-

ALGINT by James Davenport extends the integrator to expressions involving square roots of *algebraic* expressions and square roots.

ANUM by Eberhard Schrüfer extends the number domains to support algebraic numbers.

EXCALC by Eberhard Schrüfer supports exterior calculus (i.e. differential forms).

GENTRAN by Barbara Gates supports the automatic GENeration and TRANslation of program code in FORTRAN, Ratfor and C.

GROEBNER by Rüdiger Gebauer, Anthony C. Hearn, Heinz Kredel and Michael Möller supports manipulation of coupled polynomial systems.

SPDE by Fritz Schwarz finds symmetries of partial differential equations.

Of these, probably ALGINT, GENTRAN and GROEBNER are the most useful for general purposes, and so these are all that we will consider. The packages are updated periodically, but we will refer to the versions that we received with our distribution of REDUCE 3.3 in about 1987. However, a complete rewrite of GROEBNER exists in the Network Library. There is also a High Energy Physics package, HEPHYS, which is a similarly specialized package but was written by Anthony C. Hearn a long time ago, and is almost the reason for the existence of REDUCE. It is therefore sometimes built into the main REDUCE system for historical reasons, although it is quite possible to build a new version of REDUCE with a different selection of packages built into the main system. The High Energy Physics package is intended for manipulating relativistic 4-vectors, Dirac gamma matrices, etc. to support calculations in quantum electrodynamics and relativistic quantum field theory – Fitch (1985) gives a simple example of its use. (One reason for not including this package by default is that it predefines the operators `G` and `EPS`, which prevents them being used for other purposes.)

Ideally, the user-contributed packages will be available in a pre-compiled form that can be simply loaded by giving a command of the form

```
LOAD ALGINT;
```

(or possibly FLOAD in Cambridge REDUCE instead of LOAD). However, there are several reasons why this might fail. The fast-loading (FASL) files might have been put in the wrong directory – they should probably be in the main Lisp image directory used by REDUCE. If you can find them, then you should be able to load them by giving their full filestore pathnames, or you may be able to move them (or get them moved). If they do not exist then you may be able to generate them using the FASLOUT command, as described in §18.2 of the REDUCE manual (Hearn 1987) and mentioned previously in §4.3. When doing this, some of the comments below on compiling these files are relevant.

The "fast load" facility (FASLOUT and LOAD) is provided by the Lisp system and not by REDUCE, and so may not be available in all REDUCE implementations. For example, in Cambridge Lisp it has only recently become available in some versions, and is not currently available in the Atari version. If fast loading files are not available it will be necessary to input the source files (using the IN command). It is not immediately obvious from a cursory glance at these source files, but in most cases REDUCE should be put into SYMBOLIC mode before they are input. This is the case for the three packages that we will consider here. Inputting in the wrong mode may not give any clear error messages, but the package will not work[2]. They also need to be compiled in order to run at a reasonable speed.

One possibility is to read in the source each time it is required, but this is very slow. Under Cambridge Lisp a better solution (if FASLOUT is not available) is to build into the main system those packages that may be required reasonably often, which can be done using the commands

```
SYMBOLIC; ON COMP;
IN filename;
OFF COMP; ALGEBRAIC;
% Now either use the package, or
% to build it into REDUCE do:
LISP PRESERVE();
```

The precise choice of settings selected before a PRESERVE is optional, and re-selecting algebraic mode is not strictly necessary because it is reset in the REDUCE start-up code.

There are two disadvantages to compiling the packages into REDUCE (as opposed to producing fast-loading files for them). One is

[2]The suggestion made at the end of §6.10 to preface all statements by an appropriate mode declaration is intended to avoid this problem.

that some of them are mutually incompatible, e.g. EXCALC and the vector calculus package VECTOR each redefine $\wedge$ to mean respectively exterior and vector product instead of exponentiation. The other is that (at least under Cambridge Lisp) each one adds about 20Kb to the size of the basic REDUCE system, whether or not the package is in use. When the package is actually used, the amount of memory taken up is much larger. ALGINT is a particularly large package, and adds about 25Kb to the basic system. This leads us to conclude that some kind of fast load facility is almost essential to make add-on packages really usable. Nevertheless, we were able to compile all three packages and build them into REDUCE, one after the other, on a 1Mb Atari ST and still have enough memory to use them, but only for very simple problems, particularly with ALGINT. On our 1Mb Atari and Acorn machines Cambridge REDUCE loads into about 740Kb if no other facilities are also in use, and to use any of the packages it is important to keep the memory available to REDUCE as large as possible. With a 2Mb machine there should be no serious problems.

When inputting the source of the packages, and especially when compiling, one should beware that both the parser and the compiler may be upset by identifiers having "unexpected" properties. We found it necessary with all three packages to do a trial run, watching very carefully for both warning and error messages, fix any problems that showed up, and then do the final compile and preserve. In particular, the identifier `E` is flagged `CONSTANT` in REDUCE, but it is used as a procedure argument or local variable in both GROEBNER and GENTRAN. This upset the Cambridge Lisp compiler, which refused to compile the procedures affected and produced a warning (at least on the Atari), and may similarly upset other Lisp compilers. As an example of similar but local problems that might occur, in the system that we use we have defined `SAVE` to be a command for saving interactive input to a file (this is non-standard) and so the identifier `SAVE` has on its property list the property `STAT` with value `RLIS`. However, `SAVE` is used as a local variable in several procedures in ALGINT, and the `STAT` property upset the parser. The solution in all cases was to remove the offending flag or property, compile the file and then put the flag or property back, using essentially the following commands:

```
remflag('(e), 'constant);
in groebner, gentran;
flag('(e), 'constant);

remprop('save, 'stat);
in algint;
put('save, 'stat, 'rlis);
```

8.2 Integrating square roots using ALGINT

Square roots fall outside the scope of the Risch-Norman integration method (discussed in §7.2), and although the standard REDUCE integrator provides some essentially heuristic support it is very limited. ALGINT implements the restriction to square roots of an algorithm by Davenport (1981) for integrating rational powers, which is based on algebraic geometry. The use of ALGINT is controlled by the switch `ALGINT`, which is turned on by default when the package is loaded. Here is an example that shows why ALGINT is necessary:

```
OFF ALGINT;
EASY := INT(X*SQRT(1 + X**2), X);

                  2          2
          sqrt(x  + 1)*(x  + 1)
easy := -----------------------
                     3

HARD := INT(SQRT(1 + X**2), X);

                      2
hard := int(sqrt(x  + 1),x)

ON ALGINT;
REVAL HARD;  % one way to force re-evaluation

         2                      2                        2
2*sqrt(x +1)*x - log(sqrt(x +1) - x) + log(sqrt(x +1) + x)
-----------------------------------------------------------
                              4
```

Clearly, ALGINT does not use the trigonometric-type substitution that one would probably use by hand, but it is easy to show that the above integral is correct by differentiating it.

8.3 Gröbner bases for coupled polynomials

The name of Gröbner was attached to this theory by Bruno Buchberger (Professor at the University of Linz, Austria) in honour of his thesis supervisor, and much of the original work on Gröbner bases is by Buchberger. An alternative spelling of Gröbner is Groebner, which is more convenient for computer use.

Let us experiment with the package a bit before taking a look at the theory. The main procedure `GROEBNER` takes as argument a list of multivariate polynomials and tries to reduce them to an equivalent standard form (a standard or Gröbner basis) in which (with the default ordering) they are partially decoupled. The equivalence is that the result of equating the final set of polynomials (the Gröbner basis) to zero is the same as that of equating the original set of polynomials to zero. Thus, if one of the resulting polynomials is a constant (which would actually be reduced to unity) then the original polynomials have no simultaneous zero. **GROEBNER** will also accept polynomial equations, which it immediately replaces by equivalent polynomial expressions that would equate to zero.

By default, **GROEBNER** assumes that all identifiers represent variables, unless a second argument consisting of a list of variables is given (as for SOLVE). Here are some examples (with default switch settings):

```
GROEBNER {X + Y, X - Y};

{ - y,x}

GROEBNER {X + Y, X + Y + 1};

{1}

GROEBNER {X**2 - Y**2, X**2 + Y**2};

  2  2
{y ,x }

GROEBNER {X**2 + Y**2 + X, X + 2Y};

            2    4
{2*y + x,x  + ---*x}
                5
```

The last example shows the sense in which the decoupling is only partial. The second expression in the result depends only on **X** and so its zeros can be found by conventional methods. In this example they are obvious, but in general such a problem would involve a numerical solution for the roots of a polynomial that would involve real (or even complex) numbers. This is why the expressions produced by **GROEBNER** cannot in general be completely decoupled.

However, in the case that all the expressions are linear the Gröbner basis is always completely decoupled, so that extracting the zeros is trivial, and **GROEBNER** thus provides an alternative method of solving coupled linear equations as a special case of simplifying coupled polynomials, e.g.

```
GROEBNER{2X + Y + Z = 1, X + 2Y + Z = 1, X + Y + 2Z = 1};

       1       1       1
{x - ---,y - ---,z - ---}
       4       4       4
```

Coordinate geometry is a good source of nonlinear equations. For example, in a suitable coordinate system the problem of finding the intersection of a circle and a line can be expressed as

```
GROEBNER{X**2 + Y**2 = A**2,  Y = M*X + C};

     2     2     2
{ - a  + x  + y  , - c - x*m + y}
```

This output is not very helpful, and the reason is that GROEBNER did not know which were to be regarded as variables and which as (symbolic) constants. We can rectify that as follows:

```
GROEBNER({X**2 + Y**2 = A**2,  Y = M*X + C}, {X,Y});

{ - m*x + y - c,

                            2  2    2
  2       2*c               a *m  - c
 y  - ---------*y - ------------}
          2               2
         m  + 1          m  + 1
```

which has produced the desired partial decoupling. By analysing further the second basis polynomial, which is decoupled as a univariate quadratic, it would now be trivial to find, perhaps with the aid of REDUCE, geometric properties such as the conditions for the circle and line to intersect twice, touch, or not intersect, and the points of intersection.

One can invent many simple problems along these lines – as a final one where both equations are nonlinear let us consider the intersection of a rectangular hyperbola and a circle:

```
GROEBNER({X*Y = C, X**2 + Y**2 = A**2}, {X,Y});

                  2
       1  3       a
{x + ---*y  - ----*y,
       c          c

  4    2  2    2
 y  - a *y  + c }
```

which is equally easy to analyse further in this partially decoupled form.

These examples have all been intentionally trivial, and ones that could easily be done by other means, because our aim is to understand Gröbner bases. For some much less trivial problems see the documentation and test/demonstration files for the GROEBNER package. Note that for consistent expressions or equations it is always the last expression produced by `GROEBNER` (with its default switch settings, see §§8.3.1, 8.3.2) that is completely decoupled. This last one can then in principle be completely solved, each of the solutions substituted into the previous expression, it then completely solved etc. This observation generalizes to more than two variables, but in general the number of equivalent expressions produced by `GROEBNER`, and their degrees, can be high.

The procedure for solving coupled polynomial equations is analogous to that for solving coupled linear equations: the essence of any method is to reduce the system to a triangular form in which the last equation can be "trivially" solved, and then to solve the remainder of the triangular system by "back-substitution". In fact, complete Gröbner basis reduction of a linear system goes beyond triangular form all the way to diagonal form so that there is no need for back-substitution, as we have already seen. (It corresponds to Gauss-Jordan rather than just Gaussian elimination in the linear case.)

When solving coupled linear equations numerically the technique known as "pivoting" is necessary to avoid zero diagonal terms from stopping the algorithm, and small diagonal terms from causing large numerical errors. Pivoting consists of swapping rows (and sometimes even columns) so as to make the diagonal term to be used next as large in magnitude as possible. This amounts to choosing carefully the order in which to eliminate terms from the system of equations. There is analogous freedom to choose the order of elimination of terms in a Gröbner basis reduction, which is decided by an ordering on monomials (often called power products in this context). For example, should x come before y, and should x^2 come before or after y?

In fact, when a list of variables is provided to `GROEBNER` it uses them *in the order given*, and generally a different reduction will result from changing that variable ordering.

8.3.1 Some Gröbner basis theory

Our brief introduction to the theory here and in §8.3.2 and §8.3.4 is based primarily on the presentation by Davenport *et al.* (1988), but see also Buchberger (1985); both works contain many other references.

The theory revolves around the problem of solving a system of coupled polynomial equations, although it applies also to related problems.

If the polynomials $\{p_i\}$ all vanish, then so does the combination $\sum_i c_i p_i$ where the coefficients c_i are in general also polynomials. All possible such combinations generate a space called an *ideal* (introduced in §7.1). Two non-zero polynomials given by different formulae are equivalent, in that they would both give the same value at the same point, if their difference is identically zero. These ideas are formalized in the following two definitions:

Definition: The *ideal* generated by a family of polynomials consists of the set of all linear combinations of those polynomials with *polynomial* coefficients.

Definition: Two polynomials f and g are *equivalent with respect to an ideal I* if their difference belongs to I.

The ideal generated by a set of vanishing polynomials contains all information about them, and it would therefore be advantageous to find a description of the ideal that is (in some sense) minimal, i.e. a minimal set of generating polynomials or *basis* for the ideal (see also §7.1). This set of polynomials is equivalent to the original set. Defining *minimal* requires an ordering. Let $<$ be a total ordering on monomials; to be suitable for defining a Gröbner basis it must satisfy the following conditions:

- $1 < a$ for every monomial $a \neq 1$;
- if $a < b$ then $ac < bc$ for every monomial c.

A total ordering on the monomials (or power products) in each polynomial implies an ordering for the polynomials themselves. For univariate polynomials the only possible ordering is by degree, but for multivariate polynomials various choices of ordering are possible. A monomial (as defined earlier) consists of a product of variables raised to various powers, and an ordering is needed on the variable identifiers as well as on their degrees.

In a computational environment variable identifiers will frequently consist of more than a single letter. The most obvious choice of ordering on them is *lexicographic* (abbreviated to *lex*) or dictionary ordering, that is, alphabetic ordering by the first character, then the second character if any, and so on, giving $a < aa < b$. Since usually identifiers may contain characters other than letters, and upper and lower case letters may be distinct, a generalization of alphabetic order of characters is required, such as that implied by the ASCII coding sequence in which $\mathit{space} < 1 < 9 < A < Z < a < z$. Generally, the natural order of the computational system would probably be referred to as *lexicographic*.

If the sense of the above inequalities were reversed then the resulting ordering would be called *inverse lexicographic* (abbreviated to *invlex*)[3].

An ordering is also needed on variable degrees, as in the univariate case, and the freedom in the multivariate case comes from the order in which the variable and degree orderings are combined. The total degree of a monomial is the sum of the degrees of the variables, so that the total degree of the monomial xy^2z^3 is 6.

In practice, polynomials are normally written in *decreasing* total order as we will do henceforth, rather than the increasing order that we used above (as used in a dictionary). It may help to think of variables that are absent from a monomial as being present but raised to the power zero, which is how a computer implementation would probably represent such monomials in this context. Some suitable and commonly used total orderings on monomials are the following, named in terms of the dominant or first ordering, and all applied to the same polynomial $(x + y)^2 + x + y + 1$ for illustration:

- lexicographic (meaning lexicographic variable ordering followed by variable degree ordering), e.g. $y^2 + 2yx + y + x^2 + x + 1$;
- inverse lexicographic (meaning inverse lexicographic then variable degree), e.g. $x^2 + 2xy + x + y^2 + y + 1$;
- total degree (meaning total degree then lexicographic as above), e.g. $y^2 + 2yx + x^2 + y + x + 1$;
- inverse total degree (meaning total degree then inverse lexicographic as above), e.g. $x^2 + 2xy + y^2 + x + y + 1$.

In these examples, replacing the + signs by > signs shows the correct decreasing ordering.

From now on we will assume a chosen fixed ordering; the results would in general change if the ordering were changed. If a polynomial is written with its monomials in *decreasing* order then the first monomial is called the *principal monomial* and the first term is called the *principal*

[3]We have followed the definitions used by Buchberger (1985), which also appear to be those used in the distributed REDUCE GROEBNER package, but beware that Davenport *et al.* (1988) and the later REDUCE package appear to use the opposite definitions. Sit (1989) gives a careful discussion of the current confusion about term orderings in the literature, pointing out in particular that lexicographic ordering is usually interpreted in terms of a lexicographic order of the degrees of the variables.

term. (The latter is analogous to, but in general not the same as, the leading term in REDUCE terminology.)

Before we can define a minimal representation for a set of polynomials we need some criteria for discarding polynomials from the set, or at least discarding monomials. A polynomial would become "simpler", where simpler means lower down the chosen ordering, if we could remove its principal term, which we can do if it is a multiple of the principal monomial of any of the other polynomials in the set. For example, given

$$f = x^3 + y^2 \quad \text{and} \quad g = x^2 + y$$

and either inverse lexicographic or total degree ordering, we can "simplify" f by subtracting xg to give $f' = -xy + y^2$. One can produce similar simple examples where the simplified polynomial does not *look* any simpler at all, but the important point is that the procedure must be systematic. Having removed all possible principal terms, we may as well use the same process to remove any other non-principal terms possible. It can be shown that this process terminates. We can now formalize these ideas with a couple of definitions:

Definition: A polynomial f is *reduced* with respect to a finite set G of polynomials if no *principal monomial* of an element of G divides the *principal monomial* of f.

Definition: A polynomial f is *completely reduced* with respect to G if no *principal monomial* of an element of G divides *any monomial* of f.

In the case of linear expressions, removing all possible principal monomials corresponds to Gaussian elimination, and (suitably sorted) produces a triangular system. Continuing to remove all possible non-principal monomials corresponds to Gauss-Jordan elimination and reduces the triangular linear system to a completely decoupled diagonal one.

8.3.2 Standard or Gröbner bases

We now have enough background to motivate the following main definitions:

Definition: A system of generators (or a basis) G for an ideal I is a *standard* or *Gröbner* basis (with respect to an ordering $<$) if every complete reduction of an $f \in I$ (with respect to G) gives zero.

Definition: A *reduced* basis is a basis, every polynomial of which is completely reduced (as defined above) with respect to all the rest.

It can be shown that a Gröbner basis contains the same information about an ideal (or set of equations) as the original set of polynomials, although it may contain more – possibly very many more – polynomials than the original set. Some of the usefulness of Gröbner bases is contained in the following theorems:

Theorem 1: Every ideal has a Gröbner basis with respect to any ordering (as defined above).

Theorem 2: Two ideals are equal if and only if they have the same reduced Gröbner basis (with respect to the same ordering). This gives a canonical representation for ideals.

Theorem 3: A system of polynomial equations is inconsistent if and only if the corresponding Gröbner basis contains a (non-zero) constant.

Theorem 4: A system of polynomial equations has a finite number of solutions over the complex numbers if and only if each variable appears alone in one of the *principal* terms of the corresponding Gröbner basis with respect to an ordering that is firstly by identifier, e.g. lexicographic or inverse lexicographic.

An ordering firstly by variable identifier is necessary to produce the partially decoupled or triangular form that allows a system of coupled polynomial equations to be solved by back-substitution. Suppose the ordering is such that $x_1 > x_2 > \cdots > x_n$, and that the monomials within each polynomial, and the polynomials themselves, are written in decreasing order. Then if the conditions of Theorem 4 above are satisfied, the equations in Gröbner basis form must look like this:

$$\begin{aligned}
a_{11}x_1^{d_1} &+ \text{poly in } x_1 \text{ of degree} < d_1 \text{ and in } x_2, x_3, \ldots, x_n &= 0\\
a_{22}x_2^{d_2} &+ \text{poly in } x_2 \text{ of degree} < d_2 \text{ and in } x_3, x_4, \ldots, x_n &= 0\\
&\vdots\\
a_{n-1,n-1}x_{n-1}^{d_{n-1}} &+ \text{poly in } x_{n-1} \text{ of degree} < d_{n-1} \text{ and in } x_n &= 0\\
a_{nn}x_n^{d_n} &+ \text{poly in } x_n \text{ of degree} < d_n &= 0
\end{aligned}$$

Since x_n must appear alone in the principal monomial of the last equation, this equation cannot involve any other variables because they would order before x_n. Therefore the last equation must be completely decoupled, and at least if it has numerical coefficients is amenable to solution by conventional numerical methods. The next equation up can contain only x_n and x_{n-1}, and when each of the values for x_n just found are substituted into it in turn it becomes an equation in

x_{n-1} alone, which again can in principle be solved etc., up to the first equation. However, some solutions of individual equations may have to be rejected because they lead to inconsistencies in the full system.

8.3.3 The GROEBNER package revisited

The *internal* representation used by REDUCE for polynomials, as discussed in chapter 5, corresponds to none of the orderings described in §8.3.1. It is a nested representation, where within each level of nesting the monomials are ordered by degree with respect to the current main variable. The default nesting depth of variables is normally in increasing lexicographic (i.e. decreasing inverse lexicographic) identifier order, but it can be changed using the command `KORDER` (which also affects the output ordering). Therefore, one component of the GROEBNER package is devoted to converting between the standard nested representation and the fully expanded *distributive* representation (introduced in §5.1).

The default ordering of simple monomials (as distinct from terms with factors which are kernels corresponding to operator references) within polynomials output by REDUCE is normally *inverse lexicographic* as we have defined it above, but can be changed using the command `ORDER`. By contrast, the output from the GROEBNER package is ordered independently of the normal output ordering (and is in true prefix form internally). However, if it is re-evaluated in the normal REDUCE environment then the terms of each polynomial in the output list are re-ordered subject to the default variable output ordering (and converted to pseudo-prefix standard quotient form), although the order of the polynomials in the list is not changed.

The GROEBNER manual is imprecise about the term orderings used[4]. The command `TORDER` takes one of two arguments, `INVLEX` or `TOTALDEGREE`, and returns the previous ordering. We shall assume that these two modes are intended to mean the inverse lexicographic and total degree orderings that we defined in §8.3.1. The manual states that the default is `INVLEX`, which is consistent with the default ordering that REDUCE normally uses in other contexts. But the manual also states earlier that:-

GROEBNER calculates the Groebner Basis of the given set of expressions with respect to the given set of variables in the order given. If the variable list is omitted, the variables in the expression list are used, ordered according to the system variable order.

---

[4]So is the literature generally – see Sit (1989). However, the documentation with the later package is much clearer, and is consistent with our deductions about the distributed version.

Clearly the latter ordering need not be *invlex* as we have defined it, and the astute reader will have noticed an inconsistency among the orderings of the results in the examples that we showed earlier and those in the GROEBNER manual (which produce exactly the same results when run on the Atari that we used for our own examples). After experimenting and studying the code[5] we believe that the variable ordering really used is as follows. When a list of variables is supplied, GROEBNER consistently regards that list as specifying a decreasing variable order, and if the list were in alphabetic order then this would correspond precisely to decreasing inverse lexicographic order. The inconsistency occurs when no variable list is specified. The only such example in the manual corresponds to *invlex*, and our examples using *equations* for input (particularly the linear example) also correspond to *invlex*, but our examples using *expressions* for input all correspond to *lex* ordering. The reason for this is that `GROEBNER` runs through the input expressions or equations and extracts a list of all the variables, but it does not sort this list, and the order in which the variables are extracted can be affected by the precise form of the input. A striking example is to take our first example and run it first using expression input as before, and then again after trivially converting the expressions to equations:

```
GROEBNER {X + Y, X - Y};

{ - y,x}

GROEBNER {X + Y = 0, X - Y = 0};

{x,y}
```

We conclude that what the GROEBNER `INVLEX` mode really means is that it takes the list of variables either specified explicitly, or as extracted by `GROEBNER`, and *treats it as if it specified an inverse lexicographic variable ordering.* Since it is hard to predict precisely what variable ordering will be used by default, we recommend that if it matters then a list of variables should always be specified explicitly. The

---

[5]When reading the code it may help to know that degree comparisons are made using the relational operator `#<`, which is defined in the ARITH source file to be a `NEWTOK` representing `ILESSP`, which by default is itself a `SMACRO` that expands to `<`. However, if the underlying Lisp supports fast arithmetic for small integers then `ILESSP` will have been defined differently in the system-dependent "back-end" source file REND, so as to override the default. For example, Cambridge Lisp supports fast arithmetic for integers whose magnitude fits into 24 bits.

output from `GROEBNER` will always be sorted according to the currently selected ordering (although the terms of each polynomial may not remain sorted after re-evaluation by REDUCE, as we explained earlier).

This package can operate in any of the number domains supported by REDUCE, and different domains will produce different reductions. In the original GROEBNER package no facility is provided to perform the back-substitution, and only `SOLVE` is available to solve the univariate equations generated, which will work only in simple cases. The later package provides more complete support for equation solving, but still ultimately relies on `SOLVE`.

One final related task that the GROEBNER package supports is simplifying a polynomial expression subject to a set of polynomial constraints or identities, which can be performed by the techniques sketched above. The constraints or identities should first be reduced to a Gröbner basis. There are two procedures for doing this:

```
GREDUCE(exp, {exp1, exp2, ..., expn})
PREDUCE(exp, {exp1, exp2, ..., expn})
```

which both reduce *exp* with respect to the set of expressions or equations in the list. The difference is that `GREDUCE` first generates a Gröbner basis from the list, and generates an error if they are inconsistent, whereas `PREDUCE` does not. The main use of `PREDUCE` is when the list of expressions has been generated by a previous call of `GROEBNER`.

The later GROEBNER package is essentially a superset of the earlier one with the term ordering "cleaned up" and using newer algorithms. There are minor incompatibilities between the two packages, but most of our discussion applies also to the later one. We understand that a further development of the later package will be distributed with the next version of REDUCE.

The GROEBNER package supports some switches to provide further control and tracing, which we will not discuss here, but in order to understand their effects one needs to know a little more of the theory underlying the package, with which we will finish this section.

8.3.4 Buchberger's algorithm

The problem is to determine whether a basis is Gröbner or not, and if not to reduce it to one that is. The algorithm is expressed in terms of so-called H- and S-polynomials, to which the GROEBNER package switches also refer.

Definition: Let f and g be two non-zero polynomials with principal terms f_p and g_p. Their H-polynomial h is defined to be the least common multiple (lcm) of f_p and g_p. Then the S-polynomial of f and g is defined by

$$S(f,g) = \frac{h}{f_p}f - \frac{h}{g_p}g.$$

Thus, the principal terms of f and g are cancelled in $S(f,g)$. This leads to a test for whether a basis is Gröbner, as expressed in the following theorem:

Theorem: (Buchberger) A basis G is a Gröbner basis if and only if, for every pair of polynomials f and g in G, $S(f,g)$ reduces to zero with respect to G.

If any S-polynomial does not reduce to zero, then it is included in G and the test repeated. This is the essence of the algorithm used to generate a Gröbner basis, which is due to Buchberger, who has proved that it terminates. The main problem is its *computational complexity*, which is still a subject of research; in general it requires memory that increases *exponentially* with the number of variables. The process is related to that of finding greatest common divisors (gcds) of polynomials.

8.4 The GENTRAN package

The REDUCE code necessary to generate even the trivial example of a FORTRAN subprogram considered in the next chapter is quite convoluted, and larger examples would probably involve preparing most of the FORTRAN in a separate file, and grafting the REDUCE output into it by hand. The GENTRAN package is designed to support these kinds of operations more smoothly, and will produce output in FORTRAN, C or Ratfor (FORTRAN in a C-like syntax). For example, when generating C output it automatically turns each exponentiation (raising to a power) into a call to the C library function `pow`[6], which would be awkward to achieve by editing FORTRAN or "`OFF NAT`" output.

We were able to compile GENTRAN with no significant problems. The source contains all the necessary `SYMBOLIC` declarations at the head of each module, although inputting it leaves the system in symbolic mode. In addition to the problem mentioned above about `E` being declared constant, Cambridge REDUCE generates warnings about

[6] Actually, the version we have generates the function name `power`, but that is easily changed.

attempts to declare `!$EOF!$` and `!$EOL!$` (the end-of-file and end-of-line markers) to be global since they too are already declared constant. However, this did not stop the compilation, and we believe these warnings can be ignored. But we have not yet succeeded in getting the whole of GENTRAN to work properly, and we suspect that it is making assumptions about the underlying Lisp that are not true for Cambridge Lisp, although we have not yet investigated this.

In view of these problems, and the fact that the GENTRAN manual is nearly as long as the REDUCE manual itself, we will just outline its main capabilities. GENTRAN can translate REDUCE constructs into the target language, e.g. REDUCE `FOR`-loops into FORTRAN `DO`-loops. Segments of code can be evaluated by REDUCE before output in a very flexible way, and there is a mechanism for generating type declarations. GENTRAN provides a variant of the REDUCE `IN` command called `GENTRANIN`, which inputs a *template* file. This can contain text in the target language that is immediately output, plus statements that are first processed by REDUCE. The input and output handling facilities are more sophisticated than those in REDUCE itself, and allow output to multiple files simultaneously. There are powerful facilities for segmenting large expressions, and there is an interface to a target code optimizer (called SCOPE) that has recently become available in the Network Library and is claimed to produce better optimization than do optimising compilers.

...

The rest of this chapter sketches user-contributed packages (which are not currently distributed) for performing vector calculus, solving ordinary differential equations and manipulating univariate truncated power series.

8.5 Vector calculus

A vector calculus package called VECTOR has been written by David Harper at the University of Liverpool[7] (with a very small addendum by MAHM) (Harper 1989). It provides all the usual functions of Euclidean three-dimensional vector algebra and calculus (though it is restricted to orthogonal coordinate systems – for more general coordinates Schrüfer's EXCALC, described in Dr. McCrea's lectures in volume 2 of this series, could be used). The implementation defines a

[7]Harper is now at Queen Mary and Westfield College.

new REDUCE data type !3vector. The corresponding RVALUE entries are Lisp vectors with three elements. To implement functions on this data-type, VECTOR alters the definitions of the REDUCE functions GETRTYPE (to add the new type) and VARPRI (to get the correct printing mechanism) and, in order to make the scalar product an infix operator invoked by the familiar dot product notation, alters RCONS, the REDUCE interpretation of CONS in ALGEBRAIC mode.

Vectors are assigned to variables using AVEC in much the same way that matrices are assigned by MAT. A new predicate VECP is defined which AEVAL uses to test for the type !3vector and if this is found the routine VECSM!* is called to perform simplification. As well as testing directly for vectors as data, VECP tests the CAR of an expression to see if it is flagged in one of three ways, as a test for vector operators. Operators which may have vector arguments are flagged VECTORFN and have the corresponding procedure held as the VECTORFUNCTION property, scalar functions which can be applied component by component (such as DF or SIN) are flagged VECTORMAPPING, and operators taking a scalar argument but returning a vector result are flagged VECTORMAPPING and have a VECTORFUNCTION property.

The vector functions provided are CROSS, CURL, DELSQ, DIV, DOT, GRAD and VMOD (modulus of a vector). The synonyms "." and "∧" are provided for the infix operators DOT and CROSS. Vector algebra is of course provided, as is multiplication of a vector by a matrix, and the functions DF, INT, and SUB are extended to vectors. Line, surface and volume integrals are provided (which work under the same restrictions as the DEFINT definitions discussed earlier): the line integrals may be definite or indefinite. A vector substitution VECSUB is provided which enables one to write a vector argument in one set of coordinates and do the integrals in another, or with respect to parameters. With all these functions the syntax is the obvious one, e.g.

```
CURL(A);
```

where A is a vector returns the curl of A as a vector.

A similar package called ORTHOVEC, written by James Eastwood of the Culham Laboratory, Oxford, U.K., has recently been updated to version 2 and installed in the REDUCE network library.

8.6 Solving ordinary differential equations

REDUCE does not yet have a really up-to-date set of routines for solving ordinary differential equations: although many authors have implemented substantial programs these do not inter-communicate and hence do not offer the functionality of (say) MACSYMA's ODE solver, at present. MAHM has written a rather naïve program to provide a solver at the level of a British first-year undergraduate. This works by analysing the equation to find its order and whether it is linear. First order equations of the simplest standard types (e.g. Bernoulli equations) are recognized and solved, as are linear equations of higher orders with constant coefficients, and equidimensional (also called Euler) equations. Other types are just returned with an error message.

The implementation is largely in algebraic mode and done in a fairly straightforward manner, but the parsing of the equation to get its order and linearity is done in symbolic mode, and a few fixes to the way REDUCE works were also needed. For example, a new predicate `AFREEOF`[8] was defined for expressions "apparently free of" a variable, which is needed since `FREEOF` will not be true for dependent variables if tested for the independent variables. One mechanism that could usefully be added to REDUCE is a simple way to temporarily alter a switch setting inside a program at the algebraic level (see also §4.15).

The present solver is called in the form

```
ODESOLVE(ode,y,x);
```

where `ode` is the differential equation, `y` is the dependent variable to be solved for, and `x` is the independent variable.

There are ambitious plans to outdo the existing solvers in other systems by a combined effort of several groups, including the author of `SPDE` and the authors of the power-series solver DESIR (described in Davenport *et al.* 1988). This will involve using methods such as the Prelle-Singer algorithm (Prelle and Singer 1983) for first-order equations, the Lie group methods (Olver 1986, Stephani 1989) for non-linear higher-order equations and the methods of differential Galois theory for linear cases (Singer 1989). A first description of the existing and projected solvers is given elsewhere (MacCallum 1989).

8.7 Truncated power series

Alan Barnes of the University of Aston has recently produced a power series package (based on some earlier work by Julian Padget at the University of Bath) called TPS. It has a fairly small number of basic commands and introduces another new REDUCE type. This has as its `RVALUE` a rather complicated structure in which are stored the values of the coefficients of the series calculated so far and the recurrence

---

[8] A version of this is provided in the Appendix, §D.

relation for generating new terms. The simplest command is PS which generates a power series: its arguments are the function, the variable in which to expand, and the point about which to expand. For example,

```
PS(SIN X, X, 0);
```

The maximum degree is set by the PSEXPLIM command (and is by default 6): the only argument of PSEXPLIM is the new degree limit. Facilities exist to extend the usual arithmetic operations to power series, and provide composition and inversion of series, using PSCOMPOSE and PSREVERSE. PSCOMPOSE takes two arguments, each of which must be a power series, and substitutes the second in the first. PSREVERSE takes two arguments, the first being the power series and the second being the name of the expansion variable in the inverted series. (Expansion about ∞, which is given the REDUCE name INFINITY, is recognized as implying expansion in inverse powers.) The functions to be expanded can include those specified only as formal integrals or by differential equations.

PSTERM(PS,N) returns the Nth term of the power series PS, and taking PSORDER of a power series gives the degree of the first non-zero term, which may be set by PSSETORDER in cases where the package fails to do this automatically. The series may be differentiated term by term. A number of minor problems are detailed in the TPS.DOC file written by Alan Barnes.

8.8 Exercises

These exercises assume that you have access to the packages described. To conserve memory it may be best to begin each exercise with a new REDUCE session. For each exercise first load, if necessary, the appropriate package (ALGINT, GROEBNER, VECTOR, ODESOLVE, or TPS). Then run through the test file for that package and try some simple examples of your own. Use the experience gained to carry out the following calculations.

1. The Chebyshev polynomials of the first and second kind, usually denoted T_n and U_n, are orthogonal on $[-1, +1]$ with weight functions $(1-x^2)^{-1/2}$ and $(1-x^2)^{+1/2}$ respectively. Use procedure MOP from chapter 4 together with ALGINT to generate monic polynomials proportional to the first few Chebyshev polynomials. Use the alternative definitions

$$T_n(x) = \cos n(\arccos x),$$

$$U_n(x) = \sin(n+1)\theta/\sin\theta \quad \text{where} \quad x = \cos\theta,$$

to find the constant term in these polynomials, and hence generate the first few Chebyshev polynomials. [This is probably not feasible in less than about 2Mb of memory.]

2. Use the GROEBNER package to find the conditions for two general ellipses with the same centre to (a) intersect, (b) touch, (c) neither, and in cases (a) and (b) find all the points of contact. See if you can generalize your solution to the case of ellipses with different centres.

Write a back-substitution routine for use with the GROEBNER package, and test it with some simple coupled polynomial equations. [Use REDUCE itself to generate equations with known roots.]

3. (a) Integrate the divergence of the vector $r^2\mathbf{r}$ over the sphere $r = b$ and integrate the vector itself over the surface of the sphere (so verifying the Divergence Theorem).

(b) Similarly, verify Stokes' theorem for a vector $(2y, -z, 3)$ and the surface $z = 4 - r^2$ (in cylindrical polars), $z > 0$.

4. Solve the differential equations generated by equating the following expressions (all written in REDUCE form) to zero:

(a)
```
(2*X**3 - 6*X*Y + 6*X*Y**2) +
        (-3*X**2 + 6*X**2*Y - Y**3)*DF(Y,X)
```

(b)
```
(1-Z**2)*W*DF(Z,W) + (1+W**2)*Z
```
with the condition $z = 2$ at $w = \frac{1}{2}$

(c)
```
X*DF(Y,X) - Y - (X*(X+2*Y))**(1/2)
```
with the condition $y = -\frac{1}{2}$ at $x = 1$

(d) `DF(Y,X) - (X-Y)/(X+Y)`

(e) `DF(Y,X) - (X-Y-3)/(X+Y-1)`

(f) `X*(1-X**2)*DF(Y,X) + (2*X**2-1)*Y - X**3*Y**3`

(g) `DF(Y,X,4) + 8*DF(Y,X,2) + 16*Y - E**X`

(h) `X*DF(Y,X,2) + DF(Y,X) + Y/X + (LOG X)**3`

5. Find the power series expansion of $[2(u - \log(1+u))]^{1/2}$ about $u = 0$. Invert it. Write a program to repeatedly (for increasing n) increment the expansion limit to $2n+1$, compute additional terms of the inverse series and take the coefficient of its $(2n+1)^{th}$

term multiplied by the odd integers up to $2n+1$. (This gives the coefficients in Stirling's formula – a simple version, and perhaps any version, is quite slow after the first few terms: an Atari Mega ST4 took 48 hours for the first 41 odd terms.) [This problem was suggested by work of Dr. C. Tripp of Brunel University.]

9

Advanced topics

The first part of this chapter builds on chapters 5 and 6 to introduce algebraic use of symbolic mode via a case study in the development of a simple new REDUCE facility. It then discusses the use of matrices as local rather than global objects and uses some symbolic procedures to provide complete (but inelegant) support for local matrices. The second part presents an example of using REDUCE to generate FORTRAN 77 code for use in numerical computations, in which a Chebyshev-economized polynomial approximation is generated in Horner or nested form suitable for efficient numerical evaluation. Finally, a few remaining facilities are listed that have not been discussed in detail.

9.1 Mixing algebraic and symbolic modes

Most of the algebraic facilities in REDUCE are actually programmed in symbolic mode for two reasons: efficiency and flexibility. For the same reasons it may be desirable to program new facilities in symbolic mode, and moreover understanding how to do this sheds light on the implementation of the existing REDUCE facilities. After considering how arguments are passed to procedures in algebraic mode, we will consider in detail several approaches to developing a facility to apply equations as assignments.

9.1.1 Arguments to algebraic procedures

Values of local variables inside both symbolic and algebraic procedures are represented preferentially as Lisp values, and not as properties, although an algebraic value can still be extracted from the property list of an identifier. In particular, when values are passed to an algebraic procedure or an equivalent symbolic procedure from algebraic mode they are passed as Lisp values in prefix form. The easiest way to see what this means is with a small test procedure such as the following:

```
ALGEBRAIC PROCEDURE SHOW U; SYMBOLIC PRINT U$

SHOW (1/2)$

(quotient 1 2)

EQNLIST := {X = A, Y = B}$  SHOW EQNLIST$

(list (equal x a) (equal y b))

M := MAT((1,2), (3,4))$  SHOW M$

(mat (1 2) (3 4))
```

We have made `procedure show` above explicitly algebraic, and then used `SYMBOLIC` as an operator inside it to give access to the internal data representations as seen at the Lisp level. The more usual way to write this same procedure (as explained in §6.5) would be as follows:

```
symbolic operator show;
symbolic procedure show u; print u$
```

This pair of symbolic-mode statements produces *exactly* the same effect as the one algebraic-mode statement above; they both generate the same Lisp function `show` and cause the identifier `show` to be given the flag `OPFN`, which is the simplest way to make a Lisp function available in algebraic mode. We will refer to such a function as an "opfn", regardless of how it was generated. (But note that opfns are completely unrelated to the effect of algebraic operator declarations.)

As explained in §6.9, another way of seeing the Lisp form of arguments inside procedures is to *trace* the procedures (e.g. in Cambridge REDUCE `show` could be traced by `TRACE 'show` and the tracing cancelled by `UNTRACE 'show`); a Lisp backtrace enabled by `ON BACKTRACE` and caused by a Lisp-level error also shows the Lisp forms of procedure arguments.

9.1.2 Developing an ASSIGN facility

In order to use the solutions of equations returned by `SOLVE` in subsequent calculations it is necessary to "apply" the equations that `SOLVE` uses to return solutions. An equation (e.g. `a = b`) can be applied locally to an expression by making it an argument of a `SUB` operator, which replaces all occurrences of `a` in an expression by `b`. We used this technique in `procedure LCCODE` in §4.13. To make the equation apply globally it needs to be turned into an assignment (of the form `a := b`) or a let-rule (of the form `LET a = b`). However, `LET` cannot be given an argument that evaluates to an equation in the way that `SUB` can.

The basic problem is to turn the equation "`a = b`" into the assignment "`a := b`" where the original equation is the value of an algebraic variable or operator or an element of a list, e.g. notionally if "`eqn := (a = b)`" to perform "`LHS eqn := RHS eqn`". But this will not work directly, because the left side of an assignment (or a let-rule) is not evaluated. However, the algebraic operator `SET`[1] takes two arguments, assigns the value of its second argument to *the value of* its first, and returns the value. Thus `SET(LHS eqn, RHS eqn)` will do what we want. (The operator `SET` is so called by analogy with a function of the same name in Lisp that performs the analogous operation, as described in §6.1, but the algebraic-mode `SET` is not the same function as the Lisp or symbolic-mode `SET` – an important point to which we will return later.)

We now know how to apply a single equation, but a set of equations will normally be packaged as a list – `SOLVE` returns either a list of equations or a list of lists of equations. For each solution that `SOLVE` finds it returns an equation with an unknown on the left, otherwise it returns an expression on the left equated to 0. Hence before attempting to apply an equation we should check that it has an identifier on the left. The approach we will adopt is to write a procedure that examines its argument and proceeds accordingly.

The only remaining problem is how to distinguish the valid types of argument, but we saw in §9.1.1 that the internal representation of an equation is a Lisp list beginning with the atom "`equal`", whereas that of an algebraic list is a Lisp list beginning with the atom "`list`". Hence the distinction can easily be made at Lisp level, i.e. in symbolic mode. One approach is to write, along the lines of the `ODDP` predicate presented in §3.5, the following two predicates (named following C nomenclature):

```
symbolic operator isequation;
symbolic procedure isequation arg;
   eqcar(arg, 'equal);

symbolic operator islist;
symbolic procedure islist arg;
   eqcar(arg, 'list);
```

The function `EQCAR` returns true if the `CAR` of its first argument is identical to its second argument. We discuss `EQCAR` and related topics

[1]New in REDUCE 3.3.

further in the Appendix, §E. (There is a predicate called **EQEXPR** and defined in the ALG1 source file that REDUCE uses internally to check whether an expression is an equation; it is like `isequation` but it performs more checking. There does not appear to be any analogue of `islist` defined in REDUCE; essentially the same test that we have used is made "inline" wherever it is required.)

9.1.3 ASSIGN as an algebraic procedure

In terms of these two new predicates we can write the following purely algebraic procedure, which we will call **assign** after a similar facility in Maple:

```
algebraic procedure assign arg;
   if isequation arg then % SINGLE eqn - return its value
      set(lhs arg, rhs arg)
   else if islist arg then
      foreach eqn in arg do assign eqn
   else
      write "*** assign: ", arg, " ignored - invalid arg";
```

If its argument is a single equation then **assign** applies it as an assignment and returns its value; if its argument is a list then **assign** calls itself recursively on the elements of the list but returns nothing (there is no unambiguous value to return). In the event of an invalid argument we have chosen not to call **RedErr**, which would terminate the procedure immediately, but rather to output a warning message that copies the standard REDUCE style. This way all valid equations get applied, regardless of where they appear in complicated nested lists, and all invalid arguments are ignored apart from generating a warning.

The above procedure can be used in the following ways:

```
clear x;     assign(x=a);
clear x,y;   assign{x=a, y=b};
clear x;     eqn := x=a$  assign eqn;
clear x,y;   eqnlist := {x=a, y=b}$  assign eqnlist;
```

We have included the **CLEAR** statements because, unlike an assignment statement, our **assign** procedure assigns (via **SET**) to the *value* of the left side of the equation. Hence, without the clear, the second call of **assign** would actually perform the assignment "`a := a`", which will cause infinite recursion in the simplifier the next time that **a** is used. We return to this problem later.

The predicates could be embedded in `procedure assign`, thereby saving one level of function calls and gaining (marginally) some efficiency, thus:

```
algebraic procedure assign arg;
   if symbolic eqcar(arg, 'equal) then
      set(lhs arg, rhs arg)
   else if symbolic eqcar(arg, 'list) then
      foreach eqn in arg do assign eqn
   else
      write "*** assign: ", arg, " ignored - invalid arg";
```

Alternatively, REDUCE provides a mechanism for combining the readability of our first version of **assign** with the efficiency of the second via "substitution macros" called "smacros", which were introduced briefly in §6.8. Smacros are available only in symbolic mode, and they must be defined before they are used in other definitions, unlike normal procedures. Here is an implementation of **assign** using smacros:

```
symbolic smacro procedure isequation arg;
   eqcar(arg, 'equal);

symbolic smacro procedure islist arg;
   eqcar(arg, 'list);

algebraic procedure assign arg;
   if symbolic isequation arg then
      set(lhs arg, rhs arg)
   else if symbolic islist arg then
      foreach eqn in arg do assign eqn
   else
      write "*** assign: ", arg, " ignored - invalid arg";
```

We will not use smacros in subsequent versions of **assign**.

9.1.4 ASSIGN as a symbolic operator

The Lisp generated by REDUCE or RLISP input can be examined by setting the switch `ON DEFN`, after which the Lisp generated will be output but not executed (except for a few special cases, see §6.10). Alternatively, the Lisp code for a procedure called **proc** could, if it is *not* compiled, be examined in symbolic mode by `PRETTYPRINT GETD 'proc` (or just `PRETTYPRINT proc` in Cambridge Lisp). Alternatively, it could be displayed as RLISP by either `EDITDEF proc` or `SYMBOLIC RPRINT GETD 'proc`.

Such examination together with inspection of the definitions in the source code of any internal procedures called will show that algebraic-mode code tends to contain redundant simplifications and error checking. These are necessary to make algebraic mode reliable and simple to use, but they can be eliminated in the interest of efficiency in a specific application. The first step is to switch from basically-algebraic to basically-symbolic code, based on that generated by the REDUCE parser, as exemplified by the next implementation of **assign**:

```
symbolic operator assign;
symbolic procedure assign arg;
   if eqcar(arg, 'equal) then
      algebraic set(lhs arg, rhs arg)
   else if eqcar(arg, 'list) then
      foreach eqn in cdr arg do assign eqn
   else
      write "*** assign: ", arg, " ignored - invalid arg";
```

In this version, any algebraic-mode operation is made the argument of an **ALGEBRAIC** operator. Apart from this, the only difference is that if the argument is an algebraic list, then the **FOREACH** iteration is applied to its **CDR**, which is a Lisp list of the elements of the algebraic list.

The next step is to go fully symbolic, which requires a little more understanding of the internal operation of REDUCE. The main algebraic facilities are not implemented directly as "opfns" for two reasons. Firstly, their identifiers might clash with existing Lisp functions, as we have mentioned for **SET**. Secondly, an operator may wish to remain unevaluated, e.g. `DF(Y, X)` if `Y` has been declared to **DEPEND** on **X** and the dependence has not been made explicit, or **ABS** applied to a non-numerical argument. Such operators must examine their arguments, and then if necessary "return themselves". If an opfn did this it would simply call itself recursively (in an infinite loop!), which is not what is required.

The solution is that REDUCE uses the property list again, just as it does to store algebraic values so that they do not clash with Lisp values. If an identifier is used as an operator, and that identifier has a **SIMPFN** property on its property list, then the parser calls the internal procedure whose name is the value of the **SIMPFN** property, instead of the original operator identifier (cf. §6.11). We will refer to such an internal procedure as a "simpfn". For example, to find out how the algebraic **SET** operator is implemented, we take a look at its property list[2]:

```
PLIST 'SET;
((simpfn . simpset) (!:pp!-format . 1))
```

This shows that the function actually called is `SIMPSET` (defined in ALG1). All simpfns must be defined with a single argument; the algebraic processor packages the actual arguments with which the operator is called into a list, which is passed unevaluated to the simpfn.

---

[2]Remember that for Lisps other than Cambridge Lisp the function that returns a property list may be called something different, such as `PROP`.

Thus the algebraic code SET(a, b) generates the internal RLISP call SIMPSET('(a b)). By observing that the algebraic equation "a = b" is represented internally as (EQUAL a b), the CDR of which is what needs to be passed as the single argument to SIMPSET, we can streamline our **assign** procedure into the following fully-symbolic version, the remaining details of which we will explain after showing the code:

```
symbolic operator assign;
symbolic procedure assign arg;
   if eqcar(arg, 'equal) then
      % Do ALGEBRAIC set and return pseudo-prefix form
      mk!*sq simpset cdr arg
   else if eqcar(arg, 'list) then
      foreach eqn in cdr arg do assign eqn
   else lprim
      list("assign:", arg, "ignored - invalid arg");
```

Simpfns return standard quotient forms, whereas opfns return prefix forms, so that the standard quotient form returned by SIMPSET must be converted to a prefix form before it can be returned by this opfn version of **assign**. There are two ways to do this as explained in §6.4: the built-in function PREPSQ performs a genuine conversion, whereas MK!*SQ produces the pseudo-prefix form (!*SQ *std_quotient_form* T), which is recognized as a prefix form by the rest of the system. It has the advantage that it involves much less conversion effort, and if in fact the next function to operate on it requires a standard quotient form then it can be converted back very quickly (by the built-in function SIMP!*, which looks to see if a prefix form is actually a fully simplified pseudo-prefix "!*SQ" form, and if so pulls out the standard quotient form, but if not calls SIMP to do the full simplification and/or conversion).

The other change we have made is to replace the WRITE used previously to display error messages by a call of LPRIM. This is the standard RLISP warning message printer, and appears to be a mnemonic for "List PRInt Message". It accepts either an atom or a list as argument and, if the switch MSG is on (which it is by default), prints its argument suitably spaced as a warning message with the "***" prepended, or does nothing if MSG is off. It prints to the terminal in addition to any file to which output has been redirected (see §2.5).

9.1.5 ASSIGN as a "simpfn"

An increased level of flexibility would allow an arbitrary number of equations or lists of equations as arguments, not necessarily contained in algebraic lists. As we have seen, one way to achieve this is to implement **assign** as a simpfn, which in this case need not have a different name from its algebraic identifier. All we have to remember is that the argument will *always* be a (Lisp) list even if the operator was called with a single argument, and the list of arguments will be Lisp-evaluated but not REDUCE-evaluated, so we must iterate through the list of arguments and REDUCE-evaluate each one. (Of course, the latter is a design decision – we could write a version that did not evaluate its arguments, but its use would then be restricted to arguments that were explicitly either equations or lists of equations, and would not allow variables that evaluate to them.)

A simpfn always returns a value in standard quotient form and the value returned by SIMPSET is, of course, already in the right form. But if we allow more than one top-level equation, it is not clear which value to return. Let us choose to return the value of the right side of the last top-level equation; here is the code to implement this design:

```
put('assign, 'simpfn, 'assign);
symbolic procedure assign arglist;
   begin scalar val;
      foreach arg in arglist do <<
         arg := reval arg;  % REDUCE-eval the argument
         if eqcar(arg, 'equal) then
            % Do ALGEBRAIC set and return
            % standard quotient form
            val := simpset cdr arg
         else if eqcar(arg, 'list) then
            assign cdr arg
         else lprim
            list("assign:", arg, "ignored - invalid arg")
      >>;
      return val
   end;
```

(Slightly more compact code would result if the FOREACH loop were implemented as an explicit IF ... GOTO loop within the BEGIN ... END.)

9.1.6 ASSIGN as an "rlistat"

In view of the ambiguity over what value to return when more than one argument is allowed, it might be a better design never to return anything, and to regard assign as a *command* like ON and OFF rather than an operator. Such commands operate on comma-separated sequences of objects, which do not need to be enclosed in parentheses, and they never return algebraic values. They are implemented as symbolic procedures that are declared in an RLISTAT statement (which takes a Lisp list of identifiers as its argument), so we will refer to such procedures as "rlistats". The RLISTAT declaration puts the indicator STAT with property RLIS on the property list of the identifier.

Arguments are passed to rlistats by the parser in exactly the same way they are passed to simpfns. An rlistat command is available in both symbolic and algebraic modes with no further declarations, and indeed if a procedure is declared RLISTAT before it is defined, and calls itself recursively (as does our **assign** procedure), then the recursive call will be parsed as an rlistat. This can be avoided by placing the RLISTAT declaration after the procedure definition (and for complete safety the STAT property can be removed before the procedure is defined).

An rlistat is not intended to return a value (other than NIL), but if it does then it will not be interpreted as an algebraic value at top level and the internal Lisp value will be output. This is really the only difference between an rlistat and a simpfn, so our rlistat version of assign below always returns NIL, which does not generate any output at all in algebraic mode.

There are a couple of other significant enhancements in our final version. The warning message printer LPRIM is a purely symbolic-mode facility, so that in the event of an error it prints the internal form of the offending argument, rather than the algebraic form in which it was (presumably) input. There is a facility for printing algebraic-mode error and warning messages called MSGPRI, defined in the ALG1 source file, which we interface to **assign** via an internal procedure called ASSIGN!-MSGPRI. It is slightly more complicated to use MSGPRI than LPRIM, in that it takes 5 arguments that are explained as a comment below the definition of ASSIGN!-MSGPRI. MSGPRI prints to the terminal in addition to any file to which output has been redirected (see §2.5).

As we remarked earlier, evaluating a variable to which any expression has been assigned that depends on that variable causes a simplification recursion overflow. Since it can serve no useful purpose to make such assignments it seems sensible for **assign** to be able to detect and ignore them. The standard REDUCE facilities do not make such a check, perhaps to avoid the execution time penalty. We therefore introduce a switch called ASSIGNCHECK, which is off by default. If it is turned on it enables dependence checking, otherwise the checking is skipped. The SWITCH declaration is all that is needed to declare a switch and set it to be off by default; the name of the actual internal variable used to represent the switch is the name of the switch with !* prepended, and its value is either T for on or NIL for off.

The best place to perform the checking is inside SIMPSET.Partly for this reason, and partly to facilitate further streamlining, we have replaced the call of SIMPSET by inline code based on SIMPSET and SETK as defined in the ALG1 source file, which implement the algebraic SET and assignment respectively. Following the example of SET we only allow assignments to identifiers, which satisfy the IDP predicate, and not to expressions. The internal procedure that actually effects the assignment is LET2, also defined in ALG1. We have applied MK!*SQ SIMP!* to the value of the right side of the equation in order to make it into a standard quotient in pseudo-prefix form as is used by SET and assignment. Elsewhere the internal forms used are true prefix as returned by REVAL.

This is currently our preferred version of the assign facility:

```
% COMMAND to apply ARBITRARY number of equations or
% (possibly nested) lists of eqns as assignments -
% no value returned.  Can use without parentheses
% regardless of number of args.

switch assigncheck;  % off by default
% If assigncheck is on then assign disallows  x = x

remprop('assign, 'stat);  % to allow re-definition
% Will be declared an rlistat AFTER it has been defined.

symbolic procedure assign arglist;
   foreach arg in arglist do
      if eqcar(arg := reval arg, 'equal) then
      begin scalar x, u;
         % This block based on simpset & setk in ALG1
         if (u:=cdr arg) and idp(x:=car u) and (u:=cdr u)
         and (not !*assigncheck or freeof(car u, x)) then
            let2(x, mk!*sq simp!* car u, nil, t)
         else assign!-msgpri(arg)
      end
      else if eqcar(arg, 'list) then assign cdr arg
      else assign!-msgpri(arg);

% This declaration must be here, AFTER assign defined,
% because would otherwise apply to above recursive call
rlistat '(assign);
% or equivalently: put('assign, 'stat, 'rlis);

symbolic procedure assign!-msgpri(arg);
   msgpri("assign:",arg,"ignored - invalid arg",nil,nil);
   % args: message1, algexpr1, message2, algexpr2, error?
```

9.2 Local matrices

It is quite often very convenient to be able to manipulate matrices as local objects, which primarily means to be able to pass them as arguments to procedures, return them as procedure values, and to declare and manipulate them as local objects within compound statements (i.e. BEGIN ... END blocks). We will first explain how to do this, and then give some simple examples where it is useful and natural to use local matrices.

In fact, REDUCE already supports local matrices except that there is no local analogue of the **MATRIX** declaration and it is not possible to access local matrix elements, but any operations on matrices as composite objects will work. For example, one can declare a variable to be scalar within a compound statement (thereby making it local), assign a matrix to it using the **MAT** operator, manipulate it using matrix operators, and return such local matrices as parts of the value of the compound statement. Here is a silly example, a compound statement that evaluates to a list containing a matrix and a scalar, which works exactly as one would expect:

```
begin scalar m, tpm, detm;
   % Establish an arbitrary local matrix:
      m := mat((a,b),(c,d));
   % Compute its transpose as a local matrix:
      tpm := tp m;
   % and its determinant as a local scalar:
      detm := det m;
   % Return a list constructed from these local objects:
      return {m + tpm, detm}
end;
```

9.2.1 Support for local matrices

To support local matrices fully it is necessary to have local analogues of matrix indexing, and of the **MATRIX** declaration that defines a global zero matrix of a specified size. The difficulty arises because the value of a global matrix variable is stored on its property list, whereas that of a local matrix variable is stored as its Lisp value. Thus, if `M` is a global matrix then `M(2,2)` evaluates to the (2,2) element of the matrix with identifier `M`, but if `M` is a local matrix then REDUCE sees internally e.g. `(MAT (1 2) (3 4)) (2,2)`, which it does not at present understand.

The functionality required is provided by the following three symbolic procedures `GetMatEl`, `SetMatEl` and `ZeroMat` that respectively return a local matrix element, assign (indirectly) to a local matrix element, and construct a local zero matrix of specified size. Each is declared to be a symbolic operator, which makes it equivalent to an algebraic procedure and hence accessible from algebraic mode. This approach is rather inelegant in that it does not allow local matrices to be accessed in the natural way that global matrices can be. We believe that it would not be too hard to support local matrices elegantly in the same way as global matrices.

```
symbolic operator GetMatEl;
symbolic procedure GetMatEl(M, i, j);
   % Returns M(i, j) where M is a LOCAL matrix.
   % Based on getmatelem in matr.red
   % Internally, M = (mat (m11 m12 ...)(m21 m22 ...) ...)
   <<
      LocMatCheck(M, i, j, "GetMatEl");
      nth(nth(cdr M, i), j)  % element j of row i
   >>;  % GetMatEl

symbolic operator SetMatEl;
symbolic procedure SetMatEl(M, i, j, value);
   % The assignment M := SetMatEl(M, i, j, value) effects
   % M(i,j) := value where M is a LOCAL matrix.
   % Based on setmatelem in matr.red
   <<
      LocMatCheck(M, i, j, "SetMatEl");
      rplaca(pnth(nth(cdr M, i), j), value);
      M
   >>;  % SetMatEl

symbolic procedure LocMatCheck(M, i, j, Proc);
   if not eqcar(M,'mat) then  rederr list
      (Proc, ": first argument not a matrix")
   else if not fixp i then  rederr list
      (Proc, ":", i, "invalid as first matrix index")
   else if not fixp j then  rederr list
      (Proc, ":", j, "invalid as second matrix index");

symbolic operator ZeroMat;
symbolic procedure ZeroMat(rows, cols);
   % Returns a zero LOCAL matrix of given size.
   if fixp rows and fixp cols then
      'mat . for i := 1 : rows collect
             for j := 1 : cols collect 0
   else rederr "ZeroMat: invalid matrix dimensions";
```

The code in the above procedures is based on the code in the MATR source file that performs the same tasks for global matrix variables. Procedure `ZeroMat` is analogous in use to the built-in operator `MAT`, and can be used wherever `MAT` could be used. Procedures `GetMatEl` and `SetMatEl` could be applied to global matrices, but there is not much point.

The functions `NTH` and `PNTH` are defined in the ALG1 source file: `NTH` returns the nth element of a list and `PNTH` returns the list beginning at the nth element of its argument, to which `RPLACA` (see §6.2) can be applied to replace the `car`; each checks that its argument is positive, *assuming* it to be an integer. Appropriate argument checking for both `GetMatEl` and `SetMatEl` is done by the internal procedure `LocMatCheck`. Note that `SetMatEl` has to return the local matrix as its value, and this value must be assigned back to the original matrix in the calling code. This is because arguments are passed to procedures by value, so that although `SetMatEl` changes its matrix argument directly using `RPLACA` rather than constructing a copy, the argument itself is only a copy, and therefore if the changed structure is not returned and assigned `SetMatEl` has no effect.

The following algebraic procedure to return the $n \times n$ unit matrix provides a simple illustration of the use of `ZeroMat` and `SetMatEl`:

```
procedure UnitMat n;
   % Returns the  n x n  unit matrix.
   begin scalar m;  m := ZeroMat(n, n);
      for i := 1 : n do  m := SetMatEl(m, i, i, 1);
      return m
   end;
```

Another useful procedure is one to return the Hermitian conjugate of a matrix, i.e. the complex conjugate of its transpose. The following procedure does this by first transposing the matrix using the built-in operator `TP`, and then runs through the elements of the transpose accessing each using `GetMatEl`, conjugating it using `SUB` and assigning it back to the transpose using `SetMatEl`. The procedure also illustrates the use of the "`WHERE`" statement.

```
procedure HermConj M;
   % Returns Hermitian conjugate of a (complex) matrix.
   begin integer rowmax, colmax;
      M := tp M;
      << rowmax := first len; colmax := second len >>
         where len = length M;
      for row := 1 : rowmax do for col := 1 : colmax do
         M := SetMatEl(M, row, col,
                  sub(i=-i, GetMatEl(M, row, col)));
      return M
   end;
```

However, we feel obliged to remark that whilst the above version of `HermConj` has the advantage that it is purely algebraic, it has the disadvantage that it is rather inelegant and very inefficient. We could use knowledge of the internal representation of matrices to write the following more succinct version that uses a combination of algebraic and symbolic modes. This could surely be developed a lot further, in the way that we illustrated earlier with `assign`.

```
symbolic operator HermConj;
symbolic procedure HermConj M;
   'mat . foreach row in cdr algebraic tp M collect
            foreach el in row collect
               algebraic sub(i=-i, el);
```

9.2.2 Matrix norms and ill-conditioned problems

Apart from the inherent interest of this topic, it will illustrate the use of local matrices and how to pass a procedure name as an argument to another procedure.

The *norm* of a mathematical object is a non-negative real number that quantifies the magnitude of the object; for example, the usual definition of the norm of a number is its magnitude or absolute value, and the usual or Euclidean norm of a vector is its length, but other norms can be defined (see also §4.14). The norm of a square matrix can be defined relative to a vector norm, and we will consider two (non-Euclidean) vector norms that lead to matrix norms that are useful and simple to compute. These are the 1-norm, which is the maximum column sum of absolute values of the elements, and the ∞-norm, which is the maximum row sum of absolute values of the elements. Clearly the ∞-norm is the 1-norm of the transpose of the matrix. These norms can be programmed as REDUCE algebraic procedures in a natural way using `GetMatEl`:

```
procedure Norm1(M);
   % Returns 1-norm of matrix M =
   % max. column sum of absolute values.
   begin scalar norm, colsum;  integer n;
      norm := length M;  % list {nrows, ncols}
      if (n := first norm) neq second norm then
         rederr "Norm1: matrix not square";
      norm := 0;
      for j := 1 : n do <<
         colsum :=
            for i := 1 : n sum abs GetMatEl(M, i, j);
         if colsum > norm then norm := colsum
      >>;
      return norm
   end;

procedure NormInf(M);
   % Returns infinity-norm of matrix M =
   % max. row sum of absolute values.
   Norm1(tp M);
```

Matrix norms are important in numerical analysis, and can be used to assess the *condition* of a matrix problem. An ill-conditioned problem is one that is extremely sensitive to small perturbations of the kind that are unavoidable when using approximate arithmetic such as floating-point. An ill-conditioned matrix is one whose inversion, or any operation equivalent to inversion such as solving a system of linear algebraic equations, is ill conditioned.

When examining the effects of perturbations on matrix problems it becomes clear that what is important is the *condition number* defined for a matrix A as $k(\mathrm{A}) \equiv ||\mathrm{A}|| \cdot ||\mathrm{A}^{-1}||$, where $||\mathrm{A}||$ denotes a norm for the matrix A. The condition number is large for an ill-conditioned matrix. The above definition of condition number works for all possible choices of norm, and therefore so should a REDUCE implementation. Of course, the choice of norm has to be made specific before the condition number can be actually computed, so a reasonable implementation is to make the norm an argument of a condition number procedure, as follows. This works because the name of a procedure can also be used as if it were a simple variable, and the two uses can be distinguished by context.

```
procedure Condition!-Number(M, Norm);
   % Returns condition number of matrix M with
   % respect to given norm (procedure name).
   Norm(M) * Norm(M**(-1));
```

Ill-conditioned matrices are nearly singular, but not quite, so that they have a very small but non-zero determinant. It is hard to compute any property of an ill-conditioned matrix reliably using approximate arithmetic, because the problem is ill conditioned! But ill condition is not relevant to a computer algebra system, because it computes exactly (by default) so that there is no perturbation. Hence an algebra system provides a very useful tool for studying and solving ill-conditioned problems. We will finish this section with a simple example.

Ill-conditioned matrices can arise in lots of situations – one example is in attempting to perform a least-squares approximation of some function with respect to a monomial basis. (This particular problem is usually avoided by using instead an orthogonal basis.) The matrices that arise are called Hilbert matrices, and are notoriously ill conditioned. The $n \times n$ Hilbert matrix H_n has elements $H_n(i,j) = 1/(i+j-1), 1 \leq i,j \leq n$. Keeping to the spirit of using local matrices, the following procedure generates and returns as its value the Hilbert matrix H_n, and its use is illustrated by computing H_5. In order to display the matrix in a compact form we have written a short matrix printing procedure that uses knowledge of the internal form of a matrix, and calls some of the routines in the "Basic output package for symbolic expressions" that are used internally by REDUCE and are defined in the ALG2 source file. (This is likely to be unnecessary in future versions of REDUCE.)

```
procedure Make!-Hilbert(n);
   % Returns Hilbert matrix of order n.
   begin integer i1;
      scalar M;  M := ZeroMat(n, n);
      for i := 1 : n do <<
         i1 := i - 1;
         for j := 1 : n do
            M := SetMatEl(M, i, j, 1/(i1 + j))
         >>;
      return M
   end$


symbolic operator MatrixPrint;
symbolic procedure MatrixPrint M;
   % Crude first implementation of a matrix printer
   % that calls the mathprint routines defined in ALG2.
   if not eqcar(M, 'mat) then
      rederr "MatrixPrint: argument not a matrix"
   else <<
      terpri!* t;
      foreach row in cdr M do <<
         foreach el in row do
            << prin2!* "  "; maprin el >>;
         terpri!* t
         >>
   >>$


% Here is the 5 x 5 Hilbert matrix:
off ratpri;  % to save space
MATRIXPRINT(H5 := MAKE!-HILBERT 5);

  1  1/2  1/3  1/4  1/5

  1/2  1/3  1/4  1/5  1/6

  1/3  1/4  1/5  1/6  1/7

  1/4  1/5  1/6  1/7  1/8

  1/5  1/6  1/7  1/8  1/9


% Try inverting it in float mode:
ON FLOAT;

  H5**(-1);

***** singular matrix
  OFF FLOAT;
```

```
% Why is this?
DET H5;

1/266716800000

CONDITION!-NUMBER(H5, NORM1);

943656
```

The ill condition of the Hilbert matrices is shown by the fact that we cannot compute the inverse of even the 5×5 matrix using single precision (4-byte) floating point arithmetic, because during the computation to this precision the matrix becomes singular (5×5 is the smallest matrix for which this happens). If we revert to exact arithmetic and compute the determinant we see that the matrix is not really singular, although the determinant is rather small! Finally we compute the condition number, which incidentally involves computing the inverse, and is rather large. (Since the Hilbert matrices are symmetrical, the 1-norm and ∞-norm are identical, although other choices of norm would give different condition numbers of similar magnitude.)

9.3 Numerics and FORTRAN: Chebyshev economization

This section illustrates another application of REDUCE in a numerical context. It demonstrates why one might want to use REDUCE to generate programs in other languages, and introduces the standard REDUCE facilities for generating FORTRAN source code. Usually such code will involve conventional approximate floating-point numerical computation, and the reason for not performing the calculation itself in REDUCE is that REDUCE would be too slow, and that once the program has been constructed the power of REDUCE is not needed. We will focus on generating FORTRAN 77 code, for which some limited support is built into the standard REDUCE system. Programs in any other language could be generated without much difficulty, perhaps with the aid of GENTRAN.

Usually the reason for using REDUCE is to generate complicated formulae easily and accurately and insert them into a program reliably, avoiding any transcription errors. Whether or not the surrounding code is generated within REDUCE is not too important. In this example we will generate a small complete FORTRAN 77 function subprogram.

Numerical functions that are required to be evaluated frequently, such as elementary transcendental functions, need to be computed efficiently, which is often done using piecewise polynomial approximations such that each polynomial segment has a guaranteed maximum error within a chosen range. Such a polynomial can be generated by the method of *Chebyshev economization*, which relies on the fact that on the basic interval $[-1, +1]$ the Chebyshev polynomials *equi-oscillate* between $+1$ and -1. The technique is to construct an initial polynomial approximation that is sufficiently accurate, and then try to reduce its degree by subtracting multiples of Chebyshev polynomials whilst still keeping the accuracy sufficient.

This works because the normal method of generating the initial polynomial is by a Taylor expansion which has its approximation error increasing in some general sense away from the expansion point. Subtracting c times a Chebyshev polynomial from this increases the error by no more than c but moves the point at which the maximum error is attained, and typically causes the error to oscillate wildly as does the Chebyshev polynomial itself. But as long as the maximum error remains within the specified bounds, where it is attained and how it varies do not matter.

Taylor expansion and Chebyshev economization are much easier to perform using REDUCE than by hand, even for trivial low-degree examples. The following code is general, although for illustration we use only low degree and low accuracy, and approximate only the trivial exponential function. Chebyshev economization works by continually trying to reduce the degree of the approximating polynomial using the Chebyshev polynomial of that degree, so the Chebyshev polynomials are needed in *decreasing* degree order. The first procedure below generates a list of Chebyshev polynomials $T_n(x)$ in decreasing degree order from their recurrence relation $T_{n+1}(x) = 2xT_n(x) - T_{n-1}(x)$, and the second uses this list to perform the economization. One can do all this over the rationals, but there is not much point since the results are by definition approximations, and since the aim is to generate code for a numerical language, so we will use `FLOAT` mode which suffices for this problem. If we wanted to generate double precision numerical code then `BIGFLOAT` mode with a suitable precision would be necessary.

```
procedure MakeTList(n, var);
   % Returns list of Chebyshev polynomials in variable
   % var up to degree n, thus:  {Tn, ..., T1, T0}.
   begin scalar TList;  if n=0 then return {1};
      TList := {var, 1};
      for i := 2 : n do TList :=
         (2*var*first TList - second TList) . TList;
      return TList
   end$  % MakeTList
```

```
procedure Economize(Series, Var, Error);
   % Returns Chebyshev economization of Series in Var
   % up to the specified maximum economization Error.
   % Assumes a suitable polynomial coefficient domain in
   % use - rational or float OK, but default mode is not.
   begin integer n, twonm1, ratio;  scalar TList;
      n := deg(Series, Var);
      twonm1 := 2**(n-1);
      ratio := lcof(Series, Var) / twonm1;
      if (ratio > Error) then return Series;
      TList := MakeTList(n, var);
      while ( (Error := Error - ratio) >= 0 ) do <<
         Series := reduct(Series, Var) -
            ratio * reduct(first TList, Var);
         % Use of reduct necessary because cancellation
         % may not be perfect in floating point modes.
         TList := rest TList;
         twonm1 := twonm1/2;
         ratio := lcof(Series, Var) / twonm1
         >>;
      return Series
   end$  % Economize
```

```
% Generate a FORTRAN 77 function subprogram that uses an
% economized power series approximation to return exp(x)
% with a maximum absolute error of 1e-2 on [-1, 1].

% First construct a suitable Taylor polynomial:
ex := begin scalar term; term := 1;
return 1 + for n := 1 : 5 sum (term := term * x/n) end$

% On [-1, 1] this polynomial approximation has a maximum
% error of 3.78e-3.  For a maximum allowed error of 1e-2
% we can afford an economization error of up to:
ON FLOAT;  ERROR := 1e-2 - 3.78e-3;

error := 0.00622

% (Cambridge REDUCE will not accept CAPITAL E in numbers!)
% Economize the polynomial:
OFF ALLFAC;
EX := ECONOMIZE(EX, X, ERROR)$  EX;

             3               2
0.1770833*x  + 0.5416666*x  + 0.9973956*x + 0.9947916

% This formula is inefficient numerically, so put it into
% Horner's nested form by re-building it with automatic
% expansion switched off:
COEFFLIST := COEFF(EX, X);

coefflist := {0.9947916,0.9973956,0.5416666,0.1770833}
```

```
OFF EXP;  ECONEX := 0$
FOR EACH C IN REVERSE COEFFLIST DO ECONEX := ECONEX*X + C;
ECONEX;

0.1770833*(((x + 3.058823)*x + 5.632352)*x + 5.617648)

% Unfortunately, this form with the leading coefficient
% factored out is not optimal, but there seems no way to
% avoid it with the combination "ON FLOAT; OFF EXP",
% so instead we will build the polynomial symbolically:

ECONEX := 0$
D := LENGTH COEFFLIST - 1$
FOR I := D STEP -1 UNTIL 0 DO
   ECONEX := ECONEX*X + MKID(C, I);

% Output the FORTRAN code.
ON FORT;
OUT "ECONEX.FOR";  % to generate the FORTRAN in a file
WRITE "      real function econex(x)";
WRITE "      real x";
BEGIN INTEGER N;  SCALAR CN;  N := -1;
   FOR EACH CO IN COEFFLIST DO <<
      CN := MKID(C, N:=N+1);
      WRITE "      real ", CN;
      WRITE "      parameter(", CN, " = ", CO, ")" >>
END;
WRITE ECONEX := ECONEX;
WRITE "      end";
SHUT "ECONEX.FOR";
OFF FORT;
```

This outputs the following FORTRAN source code:

```
      real function econex(x)
      real x
      real c0
      parameter(c0 = 0.9947916)
      real c1
      parameter(c1 = 0.9973956)
      real c2
      parameter(c2 = 0.5416666)
      real c3
      parameter(c3 = 0.1770833)
      econex=((c2+c3*x)*x+c1)*x+c0
      end
```

If we were willing to use FORTRAN implicit type declaration then this subprogram could have been shorter and slightly easier to generate, but implicit typing is an anachronistic feature of FORTRAN that is best ignored. We have actually violated the standard by using lower case letters, but most modern FORTRAN compilers already accept them even when in strict ANSI mode, FORTRAN 90 includes lower case letters as standard, and we do not like source code that is all upper case! The number of digits in the constants is overkill for this low accuracy, but does no harm. The parameters will be textually substituted by the FORTRAN compiler, and provide a way around the fact that REDUCE insists on factoring out the leading coefficient in "`OFF EXP; ON FLOAT`" mode.

We have made essential use of the REDUCE operator `MKID`[3], which builds a new identifier by concatenating its second argument, which should evaluate to a non-negative integer (or identifier), onto its first argument, which should evaluate to an identifier. (The code for this operator was shown in §6.5.) If one needed to assign to such a constructed identifier (which we did not here) then the operator `SET` described in §9.1.2 could be used.

REDUCE supports generation of FORTRAN code via the switch `FORT`, which when turned on causes subsequent output to be in FORTRAN mode. This means that the assignment operator takes the FORTRAN form (`=`) and assignment statements are indented by the requisite 6 spaces, as shown by the output from the statement "`WRITE ECONEX := ECONEX`" above. Although not illustrated here, in this mode output is split over multiple lines following the FORTRAN conventions, and by default polynomial coefficients are made explicitly real to avoid unnecessary run-time type conversions. This can be turned off by setting `OFF PERIOD`, because of course one might want to evaluate a polynomial over the integers in FORTRAN.

9.4 What else?

There are a few REDUCE facilities that we have chosen not to discuss in detail, because we consider that they are normally only required for advanced or specialized use, but for completeness we list them here in roughly the order in which they appear in the REDUCE manual.

- The switch `TIME` and command `SHOWTIME` provide execution timing information.

- The operator `LINELENGTH` resets or displays the current output line length. Additional control over FORTRAN output is available: the maximum number of continuation lines and the maximum line length are controlled by the values of the variables `CARDNO!*` (a name that shows the vintage of REDUCE!) and

[3]New in REDUCE 3.3.

FORTWIDTH!*, both of which can be displayed or reset by the user (in either mode).

- The commands ORDER and KORDER reset respectively the output and internal kernel orderings. The predicate ORDP is true if its first argument is ordered ahead of its second. Rayna (1987, §3.16) discusses these facilities with some examples.
- The switch NERO controls the output of zero assignments.
- The operator STRUCTR displays the logical structure of a (possibly complicated) expression.
- There is a High Energy Physics package, as we mentioned earlier.
- There are a few more switches (including some undocumented ones).

9.5 Exercises

1. Write a symbolic procedure (opfn) that will convert a sum of (possibly negative) logarithms into the logarithm of the equivalent product (or quotient), and do so recursively at whatever level in the expression such a sum occurs. A symbolic procedure seems to be the only way of accessing the sign of each logarithm, and the prefix representation in which a symbolic procedure receives its arguments (and should return its result) seems to be appropriate for this manipulation.

2. Rewrite PROCEDURE NewtonN as presented in §4.14 using local matrices. Try using some symbolic-mode code to convert more efficiently between list and matrix representation, and to estimate the required precision more efficiently. Add an option to use either absolute or relative accuracy, and check that the requested error is positive and non-zero. Test your procedure on the trigonometric equations discussed in the text, and then invent some more of your own, and use REDUCE to check the solution by evaluating the equations at the roots found.

3. Write the code necessary to support the functions MAX and MIN properly, i.e. they should work in all appropriate number domains, and if there are any symbolic arguments they should be left symbolic but all numerical arguments should be reduced to a single numerical argument that is respectively their maximum or minimum. In a simpler way, the function ABS currently does

this (in REDUCE 3.3), and the code for it in the ALG1 source file provides a useful model. For further guidance, write an algebraic procedure (as invited in chapter 4, exercise 1) to find the maximum or minimum of an algebraic list of values, and look at the Lisp code generated. In order to avoid undesired recursive calls it is necessary to implement `MAX` and `MIN` as simpfns, or perhaps better as *psopfns*, which are identical to simpfns except that `SIMPFN` is replaced by `PSOPFN` and they return prefix forms that are not re-simplified rather than standard quotient forms.

4. Use REDUCE to generate part (or all) of a FORTRAN program to find to some chosen accuracy the (unique) root of the equation

$$\cos \sin x = x$$

using the Newton-Raphson algorithm starting from the initial approximation $x = 0$. Do the same thing for the pair of coupled equations considered in §4.14, and try it for some other equations of your own invention. If you can, compile and run the FORTRAN to make sure that it really works. If GENTRAN is available, try using it.

Appendix: some technical remarks

A Remarks arising from chapter 1

A.1 Running Atari REDUCE

The Atari ST supports both command-line and graphical interfaces, although the standard Atari *desktop* provides only a graphical interface. However, the standard version of REDUCE supplied for the Atari cannot be run from the standard desktop by simply opening an icon, and on an Atari with more than 1Mb of main memory, unless a very large problem is to be solved, it is sensible to run an interface program. The interface program supplied with the Cambridge Lisp on which REDUCE runs is called MENU+. This should be (although as supplied it is not) configured to provide in the TOOLS menu an entry REDUCE, which when selected with the mouse will run REDUCE with no further ado.

Command line interpreters (CLIs) and alternative graphical interfaces (desktops) are available both commercially and in the public domain. For example, the public domain CLI called Gulam (MAHM's current preference) provides an alias mechanism that allows REDUCE to be run by typing simply "`reduce`". Alternatively, using the replacement desktop NeoDesk[4] (FJW's current preference) together with a CLI (the additional NeoDesk CLI accessory is particularly convenient) it is possible to run REDUCE by simply double-clicking on a desktop icon (representing a batch file). The "shareware" replacement desktop GEMINI supports a similar facility, which we currently use on the Atari Mega 4 machines in our M.Sc. computing laboratory at QMW.

On an Atari 1040ST with only 1Mb of main memory it may be better not to waste precious memory running MENU+ or any other interface (or memory-resident programs or desk accessories); the same would apply whenever it is important to give REDUCE as much memory as possible or in the absence of a suitable shell. Then REDUCE

[4]By Gribnif Software.

should be run directly from the built-in desktop.

On the Atari, the standard Lisp core program is called LISP.TTP. When this filename or icon is opened from the desktop, a dialogue box appears into which should be typed parameters such as

```
IMAGE D:\RED33IMG
```

to select an image directory (folder) on hard disc partition D. If the image is in the same partition as LISP.TTP then the drive specifier (e.g. `D:`) can be omitted.

A.2 Bugs in Atari REDUCE

There are a few bugs in the (rather old) version of Cambridge Lisp for the Atari, some of which are inherited by REDUCE. In particular, there are bugs in the handling of floating point numbers which affect float mode in REDUCE. Most of these can be quite easily fixed – contact FJW, preferably by email, for details.

The interrupt signal to Atari REDUCE is *Control-c*, but its effect is unreliable. Used with delicacy it can have the desired effect of cancelling execution of the current REDUCE statement and returning to the next prompt, but it is quite likely to cause REDUCE to abort completely, especially if typed repeatedly or held down. There is usually some delay before it takes effect, so a quick tap followed by a pause to see if REDUCE has recognized it seems most effective.

A.3 Conversion between rational and float representations

The code that performs conversions and comparisons between numbers in rational (or integer) and float representations is in the ALG1 source file, and is controlled by the value of the `GLOBAL` symbolic-mode variable `FT!-TOLERANCE!*`. It appears that this variable should be set to the approximate relative error inherent in the floating-point representation in the underlying Lisp, which is probably the computer's standard single-precision representation. If this is a 4-byte representation then its relative error will be about 10^{-6}, which on the face of it should be the appropriate value for `FT!-TOLERANCE!*`.

However, the situation is not quite that simple, because REDUCE currently uses `FT!-TOLERANCE!*` as the *absolute* accuracy for *comparisons* with integers, in particular to decide whether a float is zero or one, and whether to replace a float by its integer part[5] if the switch `CONVERT` is on, as it is by default. (It would be more consistent to use

[5]This perhaps ought to be the *nearest* integer, rather than the integer truncation as currently used, although that would be slower.

a relative error test for comparison between floats and integers other than zero, and for the latter a smaller absolute error, possibly derived from `FT!-TOLERANCE!*`, could be used.)

REDUCE uses `FT!-TOLERANCE!*` as a *relative* error for general *conversions* between rational and float representations. The algorithm used to convert from rational to float is straightforward and constructs the fractional part of the float representation one decimal digit at a time until the relative contribution of the next digit would be less than `FT!-TOLERANCE!*`. (It is not possible to simply convert numerator and denominator to float mode and divide them because a REDUCE integer could overflow float representation.)

The algorithm for converting from float to rational representation is more interesting, and proceeds via a continued-fraction representation. For the theory of this algorithm see, for example, §2.2.4 of Akritas (1989). The straightforward approach would be to convert a float – which is really just a particular representation of a rational as a binary fraction – into exactly the same value but as a decimal fraction. However, the continued-fraction approach has the advantage that it is fast – it converges quadratically – and it does a better job of *interpreting* the value that the float is intended to represent. For example, the float value 0.6666 *might* have been genuinely intended to represent the rational number 3333/5000, but it is more likely that it was intended to represent 2/3. On the other hand, 0.6 probably should be interpreted as 3/5. The continued-fraction algorithm provides quite good interpretation if the value of the error parameter is chosen appropriately. If it is too small then rational conversions with bizarrely large numerators and denominators result from floating point rounding errors (see the example below); they are not strictly wrong, but they are not what a user probably expects and they look very strange. If the error parameter is too big then the conversion becomes unacceptably sloppy.

The conversion of float to rational representation seems to show up inappropriate settings of `FT!-TOLERANCE!*` more clearly than do any of the other conversions or comparisons. Here are a few examples using Atari REDUCE. This uses a 4-byte float representation, but as supplied the value of `FT!-TOLERANCE!*` is 10^{-12}, which one might expect to be too small. Indeed, it was partly the behaviour of Atari REDUCE that motivated this discussion. The apparently redundant operation below of multiplying a float value by 1 is necessary to force REDUCE to convert the value into the current (default) number domain. This produces a warning message, which is all that we will show after the first example, because it contains all of the interesting information:

```
% ft!-tolerance!* is 1e-12 by default on Atari
0.123*1$
*** 0.123 represented by 67381/547813
*** 0.6 represented by 6291458/10485763
*** 0.66 represented by 1081346/1638403
*** 0.666 represented by 170502/256009
*** 0.6666 represented by 293198/439841

SYMBOLIC(FT!-TOLERANCE!* := 1E-6)$
*** 0.123 represented by 123/1000
*** 0.6 represented by 3/5
*** 0.66 represented by 1081346/1638403
*** 0.666 represented by 333/500
*** 0.6666 represented by 2/3

SYMBOLIC(FT!-TOLERANCE!* := 1E-4)$
*** 0.123 represented by 23/187
*** 0.6 represented by 3/5
*** 0.66 represented by 33/50
*** 0.666 represented by 333/500
*** 0.6666 represented by 2/3
```

These results (together with some others not shown) suggest that a value for `FT!-TOLERANCE!*` between 10^{-6} and 10^{-4} is considerably better than 10^{-12}; in view of the use discussed above of `FT!-TOLERANCE!*` also as an absolute error limit, the best value is probably 10^{-6}, although this will still occasionally give strange conversions such as that shown for 0.66. On a computer with a more accurate float representation this problem becomes much less noticeable, but it will always be there.

The conclusion is that if you do not like the way that REDUCE converts between float and rational representations (having ascertained that there are no actual errors in the handling of floating-point numbers by the underlying Lisp) try changing the value of `FT!-TOLERANCE!*`, because there is no guarantee that the implementors of your system made an optimal choice. However, we understand that this is part of REDUCE that will be revised in the next version, so the behaviour may change.

B Remarks arising from chapter 2

One of the weaknesses of standard REDUCE at present is its relatively unfriendly user interface, to which quite small and simple enhancements make a big improvement. This is particularly true on microcomputers that do not have a great deal of sophisticated operating system support. Some of our own enhancements, partly to support our teaching use, are listed in §G at the end of this appendix.

B.1 Online documentation and help

The REDUCE manual file is large (over 300 Kb) and some editors either cannot cope with a file this big, or load it very slowly. Also, although the file contains only ASCII codes, the underlining of headings is currently formatted in such a way that most file display programs show only the underlining and not the text. There are various ways around this, but none of them are "standard". Future versions of the manual are likely to be formatted using LaTeX, and whether there will also be a version that can conveniently be perused and searched online remains to be seen. However, an interactive help facility under development by Jed Marti (RAND) is nearly finished and will presumably be incorporated into future versions of REDUCE.

B.2 Demonstration of the REDUCED switch

This is the program used to produce the table showing how REDUCE treats square roots and non-integer powers (although the output was subsequently edited by hand to improve the layout):

```
LET DOIT = <<
   WRITE "sqrt x", "   ", SQRT(X), "   ", SQRT(-X);
   FOREACH Q IN {1/2, Q, 1/3, 2/3} DO
      WRITE "x**", Q, "   ", X**Q, "   ", (-X)**Q;
>>;

OFF RATPRI$
X := 8A**6*B*C$
WRITE "              x = ", X, "   x = ", -X;  WRITE "";
DOIT$
ON REDUCED$  WRITE "ON REDUCED"$
DOIT$
```

C Remarks arising from chapter 3

C.1 Reading from the terminal

The following code provides a more flexible version of the built-in procedure XREAD, which will read an algebraic expression (only) from the keyboard after outputting a prompt to the screen, regardless of any I/O re-direction that may be in effect. It is based on code by A. P. Colvin (private communication), and the REDUCE source code.

```
symbolic operator aread;
symbolic procedure aread(prompt);
   begin scalar val, !*echo;  % Don't re-echo tty input
      rds nil; wrs nil;  % Switch I/O to terminal
      terpri();
```

```
        if null prompt then
           prompt := "Input: "; prin2 prompt;
        val := xread();
        if ifl!* then rds cadr ifl!*;  % Reset I/O streams
        if ofl!* then wrs cdr ofl!*;
        return val
     end;
```

The global variables IFL* and OFL* are used by the standard REDUCE file-handling procedures that are defined in the RLISP source file, and hold the following data structures, which explains the slightly different ways that they are used above:

```
ifl* = (filename filehandle curline!*)
ofl* = (filename . filehandle)
```

D Remarks arising from chapter 4

D.1 Passing arrays as arguments

The "feature" of REDUCE 3.3 that prevents array identifiers being used alone, and hence passed as arguments to procedures, can be disabled by redefining the internal procedure responsible for the error message to be the identity operation thus (Anthony C. Hearn, private communication):

```
symbolic procedure arrayeval(u,v); u;
```

D.2 Dependence predicates

The DFN example shows up a problem in the FREEOF predicate, which is that it ignores variables that have been declared to DEPEND on another variable if they appear within expressions, rather than alone as the first argument, and it ignores indirect dependence of the form "z depends on y depends on x". This is easily fixed by re-writing FREEOF, and we include a modified version here because it illustrates a number of useful ideas, particularly how essential recursion is in processing algebraic expressions. We also take this opportunity to include a version of the AFREEOF predicate that completely ignores implicit dependence, the need for which was mentioned in §8.6.

```
symbolic procedure FreeOf(Exprn, Ker);
   % Redefines the standard FreeOf predicate.
   not DependsOn(Exprn, Ker, nil);
```

```
symbolic operator AFreeOf;
symbolic procedure AFreeOf(Exprn, Ker);
   % Defines an "Apparently Free Of" predicate.
   not DependsOn(Exprn, Ker, t);

flag('(AFreeOf),'boolean);
infix AFreeOf;  precedence AFreeOf, FreeOf;


symbolic procedure DependsOn(Exprn, Ker, Explicit);
   % Determines if prefix expression Exprn
   % depends on kernel Ker at ANY level,
   % ignoring DEPENDent kernels if Explicit is true.
   % (Based on freeof in ALG2 and smember in ALG1.)
   Exprn = Ker or
   ( not Explicit  and  ImpDependsOn(Exprn, Ker) ) or
   not atom Exprn and
      % Exprn must be in prefix form, so test operands
      begin
   loop: if (Exprn := cdr Exprn) then
            if DependsOn(car Exprn, Ker, Explicit) then
               return t
            else goto loop
      end;

symbolic procedure ImpDependsOn(DepKer, IndKer);
   begin scalar dependences;
      if null (dependences := assoc(DepKer, depl!*)) then
         return nil;
   loop:
      if (dependences := cdr dependences) then
         if ((Dep = IndKer or ImpDependsOn(Dep, IndKer))
               where Dep = car dependences) then return t
         else goto loop
   end;
```

D.3 Errors in polynomial operators

There is a bug in the function DEG (at least in some implementations), to which allusion was also made in chapter 2, such that DEG(a, x) where a is independent of x returns NIL instead of 0, whereas if a is an explicit number then 0 is correctly returned. This is easily fixed by editing the source code for DEG, which is a symbolic procedure defined towards the end of the ALG2 source file, to change *all* occurrences of NIL to 0 (the calls of !*f2a can also be removed), and re-compiling. Alternatively, a "quick and dirty" fix is the following. This source code can be safely re-read without causing trouble, and it can be compiled, although it may not be quite straightforward to permanently preserve this modification into the system. It provides an example of using the functions GETD and PUTD introduced in §6.7 for getting and putting function definitions.

```
symbolic if not getd '!%deg then <<
   deg(1,'x)$  % to make sure relevant module is loaded
   putd('!%deg, car g, cdr g) where g = getd 'deg;
   % !%deg := deg;  also works in CULisp
   procedure deg(poly, x);
      % Converts nil value of degree to explicit 0.
      (if d then d else 0) where d = !%deg(poly, x);
      % !%deg(poly, x) or 0;  also works in CULisp
   >>;
```

There are some similar problems, which are equally easy to fix, with the operators `LTERM` and `REDUCT`.

An Atari-specific problem with the related operator `MAINVAR` is that it fails when compiled (only), which appears to be due to an obscure bug in the compiler that causes the first assignment in the procedure to be ignored, so that subsequent code fails. This can be fixed by declaring the argument (u) of `MAINVAR` to be fluid and then re-compiling.

E Remarks arising from chapter 9

E.1 RLISP equality predicates

In the `ASSIGN` code we made essential use of the function `EQCAR`, which is defined (in the RLISP source file) as

```
symbolic procedure eqcar(u,v); null atom u and car u eq v;
```

and is safer than just "`CAR U EQ V`". This is itself more efficient in such applications than "`CAR U = V`", because the declaration

```
newtok '((!=) equal);
```

(also in the RLISP source file) makes "=" a synonym in RLISP for the Lisp predicate `EQUAL`. As explained in §6.3, the predicate `EQ` just compares pointers, whereas `EQUAL` compares whole S-expressions (data-structures) recursively, which is necessary in some applications, but not in `ASSIGN`.

A related remark is that the RLISP inequality predicate is defined by

```
symbolic procedure u neq v; null(u=v);
```

so that "`U NEQ V`" is less efficient than "`NOT(U EQ V)`" in situations where pointer comparison is sufficient. (The above remarks do not apply to the algebraic-mode versions of these predicates, which are implemented differently, as are the algebraic and symbolic versions of many functions.)

F REDUCE network library and other electronic-mail facilities

A REDUCE version of `NetLib`, which is a system for the distribution of software by electronic mail (for background information see Dongarra and Grosse 1987), is supported by the RAND Corporation for registered REDUCE users. It contains new and updated user-contributed packages, bibliography files, examples, programming notes etc., and is worth perusing. To begin using this facility, send the single-line message

```
send index
```

(or equivalently the messages `help` or `info`) to the network address `reduce-netlib@rand.org`. The single line message can either be the subject of the message or the body. Similarly, for information about the availability of REDUCE on different computers use the instruction

```
send info-package
```

If you need to talk to a human, have problems with `NetLib`, or would like written rather than electronic information about REDUCE availability, send mail to `reduce@rand.org`.

Your local site may already have most or all of `NetLib`, or there may be another site nearer than RAND. For example, primarily for European users there are duplicate services at `reduce-netlib@can.nl` and `eLib@sc.zib-berlin.de`, and for academic users in the UK there is a `listserv` library facility that can be accessed initially by sending to `listserv@uk.ac.liverpool` a message containing the two lines

```
GET ALGEBRA HELPFILE
GET ALGEBRA FILELIST
```

Information about changes and additions to `NetLib` is sent at regular intervals to participants in the REDUCE forum. This is a general forum for discussing technical matters related to REDUCE (but bug reports should go either to the implementor of your system or to `reduce@rand.org`). If you are a REDUCE user and would like to join the forum you can send a message requesting inclusion to `reduce-forum-request@rand.org`. There are many local redistribution points for forum mail, e.g. the main UK redistribution is managed by `reduce-forum-request@uk.ac.nsf`, but a request to RAND should get forwarded as appropriate.

There are other electronic bulletin boards dealing with specific computer algebra systems, and also a general one. If you would like to be added to the UK redistribution list for the latter send a request to `symalg-request@uk.ac.qmw.maths`.

G REDUCE corrections and enhancements

Throughout the text we have remarked on a few minor aspects of REDUCE 3.3 that we consider to be either errors or infelicities. To date we have produced the following corrections or enhancements – the main enhancements are described further by Wright (1991):

- enhanced file handling to support default files, lists of default directories and extensions with input file searching, and querying before over-writing output files;
- `SAVE` commands for selectively saving interactive input to a file in a form ready to be re-input, probably after editing, which facilitates interactive program development – output may also optionally be saved;
- an enhanced `DISPLAY` command to show various system information including mode, number domain and switch settings;
- generalized `MAX` and `MIN`, and an improved `ABS`;
- support for row and column matrices as 1-index objects, and enhanced matrix-building operators;
- algebraic-mode operators `FIX`, `ROUND`, `CEILING` and `FLOOR` to convert arbitrary numbers to integers, and support for division with quotient and remainder;
- an `UNSHARE` command;
- corrections to `DEG`, `LTERM` and `REDUCT`;
- minor improvements to `SETMOD`;
- support on the Atari ST for the GEM file selector and mouse, the built-in Lisp file editor, and desk accessories.

The following are under development:

- an alternative multivariate polynomial degree cutoff facility;
- `CLOGS`, an operator to collect sums of logarithms.

Some of this code, or its equivalent, may be incorporated into future versions of REDUCE, or for REDUCE 3.3 it is available from FJW, preferably by email (send a request to **fjw@maths.qmw.ac.uk**).

Bibliography

Abramowitz, M. and Stegun, I. A. (1964). *Handbook of mathematical functions.* Applied mathematics series, **55**. National Bureau of Standards, Washington. (Reprinted 1968 by Dover Publications, New York).

Akritas, A. G. (1989). *Elements of computer algebra with applications.* Wiley-Interscience, New York.

Bronstein, M. (1990). Integration of elementary functions. *Journal of Symbolic Computation*, **9**, 117–73.

Bronstein, M., Davenport, J. H., and Trager, B. M. (1989). Symbolic integration is algorithmic. In *Computers and Mathematics 1989* (ed. E. Kaltofen). Academic Press, New York.

Buchberger, B. (1985). Gröbner bases: an algorithmic method in polynomial ideal theory. In *Progress, directions and open problems in multidimensional systems theory* (ed. N. K. Bose), pp. 184–232. Reidel, Dordrecht.

Buchberger, B. and Loos, R. (1983). Algebraic simplification. In Buchberger *et al.* (1983), pp. 11–44.

Buchberger, B., Collins, G. E., Loos, R., with Albrecht, R. (eds.) (1983). *Computer algebra: symbolic and algebraic computation.* (2nd edn). Springer-Verlag, Wien.

Davenport, J. H. (1981). *On the integration of algebraic functions.* Lecture notes in computer science, **102**. Springer-Verlag, Berlin.

Davenport, J. H., Siret, Y., and Tournier, E. (1988). *Computer algebra: systems and algorithms for algebraic computation.* Academic Press, London. (A second edition is now available.)

Davies, B. C. and Wright, F. J. (1991). Solving trigonometric equations using REDUCE. QMW Technical Report.

Dongarra, J. J. and Grosse, E. (1987). Distribution of mathematical software via electronic mail. *Communications of the ACM*, **30**, 403–7.

Edneral, V. F., Kryukov, A. P., and Rodionov, A. Ya. (1989). *The analytic programming language REDUCE.* (In Russian). Moscow University Press.

Fitch, J. P. (1981). User-based integration software. In *SYMSAC '81: the 1981 ACM symposium on symbolic and algebraic manipulation* (ed. P. S. Wang), pp. 245–48. ACM, New York.

Fitch, J. P. (1985). Solving algebraic problems with REDUCE. *Journal of Symbolic Computation*, **1**, 211–27.

Flatau, P. J., Boyd, J. P., and Cotton, W. R. (1988). *Symbolic algebra in applied mathematics and geophysical fluid dynamics – REDUCE examples.* Report, Ver. 1.1, Department of Atmospheric Science, Colorado State University, Fort Collins CO 80521, USA.

Gardin, F. and Campbell, J. A. (1983). A knowledge-based approach to user-friendliness in symbolic computing. In *EUROCAL 83* (ed. J. A. Van Hulzen). Lecture notes in computer science, **162**, pp. 267–74. Springer-Verlag, Berlin.

Geddes, K. O. and Stefanus, L. Y. (1989). On the Risch-Norman integration method and its implementation in Maple. In *Proceedings of the ACM-SIGSAM international symposium on symbolic and algebraic calculation, ISSAC-89*, pp. 212–7. ACM, New York.

Griss, M. L. and Hearn, A. C. (1981). A portable LISP compiler. *SOFTWARE – Practice and Experience*, **11**, 541–605.

Harper, D. (1989). Vector33: A REDUCE program for vector algebra and calculus in orthogonal curvilinear coordinates. *Computer Physics Communications*, **54**, 295–305.

Harper, D., Wooff, C. D., and Hodgkinson, D. E. (1991). *A guide to computer algebra systems.* John Wiley and Sons, Chichester, UK.

Hearn, A. C. (1987). *REDUCE user's manual*, Version 3.3. RAND publication CP78 (Rev. 7/87), The RAND Corporation, Santa Monica, CA 90-406-2138, USA. (Also available in a Japanese translation by Hiroshi Toshima from McGraw Hill Japan, 1988.)

Hirota, R. and Ito, M. (1989). *Introduction to REDUCE – Doing Symbolic Computation on PC.* (In Japanese). Science sha, Tokyo.

Kaltofen, E. (1983). Factorization of polynomials. In Buchberger *et al.* (1983), pp. 95–114.

Klimov, D. M. and Rudenko, V. M. (1989). *Computer algebra methods in the problems of mechanics.* (In Russian). Nauka, Moscow.

Kolchin, E. R. (1973). *Differential algebra and algebraic groups.* Academic Press, New York.

Kryukov, A. P., Rodionov, A. Ya., Taranov, A. Yu., and Shabligin, E. M. (1991). *Programming in the RLISP language.* (In Russian). Radio and Communications, Moscow. (In press.)

Lipson, J. D. (1981). *Elements of algebra and algebraic computing.* Addison-Wesley, Redwood City, California.

MacCallum, M. A. H. (1989). An ordinary differential solver for REDUCE. In *Symbolic and algebraic computation: international symposium ISSAC-88* (ed. P. Gianni). Lecture notes in computer science, **358**, pp. 196–205. Springer-Verlag, Berlin.

Marti, J., Hearn, A. C., Griss M. L., and Griss, C. (1979). Standard LISP report. SIGPLAN notices, **14**, (10), 48–68. ACM, New York. (Distributed in machine-readable form with the REDUCE sources.)

Metcalf, M. and Reid, J. (1990). *Fortran 90 explained.* Oxford University Press.

Moses, J. and Yun, D. Y. Y. (1973). The EZGCD algorithm. Proceedings of the 1973 ACM National Conference, pp. 159–166. ACM, New York.

Nakamura H. and Matsui, S. (1989). *Symbolic Computation in Structural Mechanics using REDUCE.* (In Japanese). Gihodo Shuppan, Tokyo.

Ochiai, M. and Nagatomo, K. (1990). *Linear Algebra using REDUCE.* (In Japanese). Kindai Kagaku sha, Tokyo.

Olver, P. J. (1986). *Applications of Lie groups to differential equations.* Springer-Verlag, New York.

Pearce, P. D. and Hicks, R. J. (1981). In *SYMSAC '81: the 1981 ACM symposium on symbolic and algebraic manipulation* (ed. P. S. Wang). ACM, New York. p 131.

Prelle, M. J. and Singer, M. F. (1983). Elementary first integrals of differential equations. *Transactions of the American Mathematical Society*, **279**, 215–29.

Rayna, G. (1987). *REDUCE: software for algebraic computation.* Springer-Verlag, New York.

Risch, R. H. (1969). The problem of integration in finite terms. *Transactions of the American Mathematical Society*, **139**, 167–89.

Schwarz, S. (1956). On the reducibility of polynomials over a finite field. *Quarterly Journal of Mathematics, Oxford*, **7**, (2), 110–24.

Singer, M. F. (1989). An outline of differential Galois theory. In *Computer algebra and differential equations* (ed. E. Tournier). pp. 3–57. Academic Press, London.

Sit, W. Y. (1989). Some comments on term-ordering in Gröbner basis computations. SIGSAM Bulletin, **23**, (2), 34–8. ACM, New York.

Stephani, H. (1989). *Differential equations – their solution using symmetry* (ed. M. A. H. MacCallum). Cambridge University Press.

van Hulzen, J. A. (1989). *Formule manipulatie m.b.v. REDUCE.* (In Dutch). Twente University, Enschede.

Wang, P. S. (1978). An improved multivariate polynomial factoring algorithm. *Mathematical Computing*, **32**, 1215–31.

Winkelmann, V. and Hehl, F. W. (1988). REDUCE for beginners: six lectures on the application of computer-algebra (CA). In *Computer simulation and computer algebra: lectures for beginners.* (ed. D. Stauffer, F. W. Hehl, V. Winkelmann and J. G. Zabolitzky). pp. 83–151. Springer-Verlag, Berlin.

Winston, P. H. and Horn, B. K. P. (1988). *LISP.* (3rd edn). Addison-Wesley, Reading, Mass.

Wright, F. J. (1991). Some portable enhancements to the REDUCE user interface. QMW Technical Report.

Participants at the School

- AGUILAR, Antonio A., Department of Physics, P. O. Box 55–534, CP 09340 México – D. F., México.
- ALVARADO, Adrian G., Universidad de Costa Rica, Escuela de Física, Laboratorio de Física Computacional, San José – Costa Rica.
- ÅMAN, Jan E., University of Stockholm, Institute of Theoretical Physics, Vanadisvägen 9, S–11346 Stockholm, Sweden. Email: `ja@sesuf51 (BITNET)`.
- ANDRADE, Evandro P., Empresa Brasileira de Telecomunicações – EMBRATEL, Av. Presidente Vargas 1012/609, 20071 Rio de Janeiro – RJ, Brasil.
- BESSA, Ivan G., Pólo Petroquímico de Camaçari, COPENE – Petroquímica do Nordeste S. A., Rua Eteno 1561, 42810 CEDEN – DITEC Bahia, Brasil.
- BONHAM, Sirley M., Centro Brasileiro de Pesquisas Físicas, Departamento de Partículas e Campos, Rua Dr. Xavier Sigaud 150, 22290 Rio de Janeiro – RJ, Brasil. Email: `marques@fnal (BITNET)` and `smb@lncc (BITNET)`.
- BRUNELLI, José Carlos, Universidade de São Paulo, Instituto de Física, Caixa Postal 20516-0100, São Paulo – SP, Brasil.
- CABRAL, Sergio C., Instituto de Engenharia Nuclear – CNEN, Cidade Universitária - Ilha do Fundão, Caixa Postal 2186, 20001 Rio de Janeiro – RJ, Brasil.
- CAMARGO, Maria Angelica O., Universidade Estadual de Londrina, Campus Universitário, Caixa Postal 6001, Londrina – PR, Brasil.

- CARDOSO, Marcio José E. M., Departamento de Físico-Química, IQ–UFRJ, Cidade Universitária – CT Bloco A-Sala 408, 21910 Rio de Janeiro – RJ, Brasil.
  Email: iqg01001@ufrj (BITNET).

- CARVALHO FILHO, Joel C., Departamento de Física, Universidade Federal do Rio Grande do Norte, Campus Universitário, Caixa Postal 1661, 59000 Natal – RN, Brasil.

- CARVALHO, Maria Auxiliadora P., Pólo Petroquímico de Camaçari, COPENE – Petroquímica do Nordeste S.A., Rua Eteno 1561, 42810 CEDEN – DITEC Bahia, Brasil.

- CASCON, Ana, Laboratório Nacional de Computação Científica, Rua Lauro Müller 455, 22290 Rio de Janeiro – RJ, Brasil.
  Email: aaaa@lncc (BITNET).

- CASTIER, Marcelo, Programa de Engenharia Química – COPPE, Caixa Postal 68502, 21945 Rio de Janeiro – RJ, Brasil.
  Email: coq99046@ufrj (BITNET).

- CHAVES, Fernando Miguel P., Universidade de São Paulo, Instituto de Física, Departamento de Física Matemática, 01498 São Paulo – SP, Brasil.

- COELHO, Haydée P., Departamento de Matemática, Universidade de Brasília, 70910 Brasília – DF, Brasil.

- COLÓN, Elio R., University of Puerto Rico at Humacao, Department of Physics, Humacao – Puerto Rico 00661.

- COSTA, Herbert F. M., Universidade Federal de Pernambuco, Departamento de Química Fundamental, Cidade Universitária, 50739 Recife – PE, Brasil.

- COUTINHO, Severino C., Universidade Federal do Rio de Janeiro, Departamento de Matemática Aplicada, Caixa Postal 68530, 21945 Rio de Janeiro – RJ, Brasil.

- CUNHA, Antonio M. N., Centro de Pesquisa e Desenvolvimento para Seguro das Comunicações – CEPESC, Caixa Postal 07/0359, Brasília – DF, Brasil.

- DEVITT, J. Stan, Department of Mathematics, University of Saskatchewan, Saskatoon, Canada S7N 0W0.
  Email: devitt@sask (BITNET).

- DIAS, Luiz Alberto V., Instituto Nacional de Pesquisas Espaciais - INPE, Av. dos Astronautas 1758, Caixa Postal 515, Jardim da Granja, 12201 São José dos Campos - SP, Brasil.

- FEE, Greg J., Symbolic Computation Group, Department of Computer Science, University of Waterloo, Waterloo, Ontario, Canada N2L 3G1.
  Email: gjfee@daisy.waterloo.edu (Internet).

- FERRARIS, Marco, University of Cagliari, Department of Mathematics, Via Ospedale 72, 09124 - Cagliari, Italy.
  Email: ferraris@vaxca2.mrgate.mr.infn.it (Internet).

- FONSECA NETO, Joel B., Universidade Federal da Paraíba, Departamento de Física, Campus Universitário, 58000 João Pessoa - PB, Brasil.

- FREIRE, Denise G., CEPEL - Cidade Universitária, Caixa Postal 2754, 20001 Rio de Janeiro - RJ, Brasil.

- GOEDERT, João, Universidade Federal do Rio Grande do Sul, Instituto de Física, Caixa Postal 15051, 91500 Porto Alegre - RS, Brasil. Email: goedert@ufrgsif.ansp.br (BITNET).

- JAKUBI, Alejandro, Departamento de Física, Pabellón I, Ciudad Universitaria, 1428 Buenos Aires, Argentina.
  Email: jakubi@rutgers.edu (Internet).

- JENKS, Richard D., Thomas J. Watson Research Center - IBM, P. O. Box 218, Yorktown Heights, New York 10598, USA.
  Email: jenks@yktvmx (BITNET).

- JOFFILY, Sergio, Centro Brasileiro de Pesquisas Físicas, Rua Dr. Xavier Sigaud 150, 22290 Rio de Janeiro - RJ, Brasil.

- KLEIN, Telma S., Laboratório Nacional de Computação Científica, Rua Lauro Müller 445, 22290 Rio de Janeiro - RJ, Brasil.

- MacCALLUM, Malcolm A. H., School of Mathematical Sciences, Queen Mary and Westfield College, Mile End Road, London E1 4NS - UK. Email: mm@maths.qmw.ac.uk (JANET).

- McCREA, J. Dermot, Department of Mathematical Physics, University College, Belfield, Dublin 4, Ireland.
  Email: mccread@irlearn (BITNET).

- McLENAGHAN, Raymond G., Faculty of Mathematics, Department of Applied Mathematics, University of Waterloo, Waterloo, Ontario, Canada N2L 3G1.
  Email: rgmclenaghan@daisy.waterloo.edu (Internet).

- MOUSSIAUX, Alain G. E., Facultés Universitaires de Namur, 61 Rue de Bruxelles, 5000 - Namur, Belgium.

- ONODY, Roberto N., Instituto de Física e Química de S. Carlos, Universidade de São Paulo, Caixa Postal 369, 13560 São Carlos – SP, Brasil.

- PAIVA, Filipe M., Centro Brasileiro de Pesquisas Físicas, Rua Dr. Xavier Sigaud 150, 22290 Rio de Janeiro – RJ, Brasil. Email: userfmp@lncc (BITNET) and fmp@maths.qmw.ac.uk (JANET).

- PARDO, Francisco, Instituto de Ciencias Nucleares UNAM, Apartado Postal 70-543, Ciudad Universitaria, México.
  Email: frapar@unamvm1 (BITNET).

- PETEAN, Silvia, Departamento de Raios Cósmicos, Instituto de Física Gleb Wataghin, UNICAMP, Caixa Postal 6165, 13081 Campinas – SP, Brasil.

- PIMENTEL, Luiz O., Departamento de Física, P. O. Box 55-534, CP 09340 México – DF, México.

- PORTUGAL, Renato, Centro Brasileiro de Pesquisas Físicas, Rua Dr. Xavier Sigaud 150, 22290 Rio de Janeiro – RJ, Brasil.
  Email: repo@lncc (BITNET).

- REBOUÇAS, Marcelo J., Centro Brasileiro de Pesquisas Físicas, Rua Dr. Xavier Sigaud 150, 22290 Rio de Janeiro – RJ, Brasil.
  Email: mjr@lncc (BITNET).

- ROQUE, Waldir L., Departamento de Matemática, Universidade de Brasília, 70910 Brasília – DF, Brasil.
  Email: roq@lncc (BITNET).

- SANTOS, Raul, Comissão Nacional de Energia Nuclear, Rua General Severiano 90 - 3º andar, 22294 Rio de Janeiro – RJ, Brasil.

- SANTOS, Renato P., Centro Brasileiro de Pesquisas Físicas, Rua Dr. Xavier Sigaud 150, 22290 Rio de Janeiro – RJ, Brasil.
  Email: rps@lncc (BITNET).

- SASSE, Fernando D., Centro Brasileiro de Pesquisas Físicas, Rua Dr. Xavier Sigaud 150, 22290 Rio de Janeiro – RJ, Brasil. Email: sase@lncc (BITNET).

- SERCONEK, Shirley, CEPESC – Centro de Pesquisa e Desenvolvimento para a Segurança das Comunicações, Caixa Postal 07/0359, 71600 Brasília – DF, Brasil.

- SILVA, Adriana S., COPPE - Universidade Federal do Rio de Janeiro, Programa de Engenharia Química, Centro de Tecnologia, Bloco G - sl. G 115, Rio de Janeiro – RJ, Brasil. Email: coq@99046@ufrj (BITNET).

- SILVA JR., Waldemar M., Instituto de Física, Universidade Federal Fluminense, Outeiro de São João Batista s/n - Centro, 24020 Niterói – RJ, Brasil.

- SIMIS, Aron, Universidade Federal da Bahia, Departamento de Matemática, Av. Ademar de Barros - Campos de Ondina, 40210 Salvador – BA, Brasil.

- SKEA, James E. F., School of Mathematical Sciences, Queen Mary and Westfield College, Mile End Road, London E1 4NS – UK. Email: jimsk@maths.qmw.ac.uk (JANET).

- SPIRA, Michel, Departamento de Matemática, Universidade Federal de Minas Gerais, Caixa Postal 702, 30161 Belo Horizonte – MG, Brasil.

- TANURE, Humberto S. R. Centro de Materiais Refratários, Fundação de Tecnologia Industrial, Pólo Urbo Industrial s/n, 12600 Lorena – SP, Brasil.

- TEIXEIRA, Antonio F. F., Centro Brasileiro de Pesquisas Físicas, Rua Dr. Xavier Sigaud 150, 22290 Rio de Janeiro – RJ, Brasil.

- TOMBAL, Philippe J. A., Facultés N. D. de la Paix, Rue de Bruxelles 61, B - 5000 Namur, Belgium. Email: phtombal@bnandp51 (BITNET).

- TOMIMURA, Nazira, Instituto de Física, Universidade Federal Fluminense, Outeiro de São João Batista s/n - Centro, 24020 Niteroi – RJ, Brasil.

- TRAVASSOS, Joel M., Observatório Nacional – DGE, Rua General Bruce 586, 20921 Rio de Janeiro – RJ, Brasil.

- WINTER, Othon C., Departamento de Matemática – UNESP, Av. Ariberto Pereira da Cunha 333, 12500 Guaratinguetá – SP, Brasil.
- WRIGHT, Francis J., School of Mathematical Sciences, Queen Mary and Westfield College, Mile End Road, London E1 4NS – UK. Email: fjw@maths.qmw.ac.uk (JANET).
- XANTHOPOULOS, Basilis, Department of Physics, University of Crete, Iraklion - Crete, Greece.
- ZAMORANO, Nelson, Departamento de Física, Universidad de Chile, Casilla 487-3, Santiago, Chile. Email: nzamoran@uchcecvm (BITNET).
- ZANCHIN, Vilson T., Universidade de Campinas – UNICAMP/IMECC, Departamento de Matemática Aplicada, 13083 Campinas – SP, Brasil.

Index